Advances in Solid and Hazardous Waste Management

Sudha Goel
Editor

Advances in Solid and Hazardous Waste Management

Second Edition

Springer

Editor
Sudha Goel
Civil Engineering Department and School
of Environmental Science and Engineering
Indian Institute of Technology Kharagpur
Kharagpur, India

ISBN 978-3-031-49143-6 ISBN 978-3-031-49144-3 (eBook)
https://doi.org/10.1007/978-3-031-49144-3

Jointly published with Capital Publishing Company
The print edition is not for sale in Afghanistan, Bangladesh, Bhutan, India, Sri Lanka, the Maldives, Nepal and Pakistan. Customers from Afghanistan, Bangladesh, Bhutan, India, Sri Lanka, the Maldives, Nepal and Pakistan please order the print book from: Capital Publishing Company.

This Springer imprint is published by the registered company Springer Nature Switzerland AG
The registered company address is: Gewerbestrasse 11, 6330 Cham, Switzerland

Paper in this product is recyclable.

Preface to the Second Edition

After the moderate success of the first edition of this book, it was only natural to continue adding more material to it and we can now present the second edition of *Advances in Solid and Hazardous Waste Management.*

The objective of this book is to provide the reader a review of major issues in solid and hazardous waste management that are not covered in standard textbooks. Several examples and case studies are provided to demonstrate the applicability of new concepts and ideas.

Key features of the book include:

- Remote sensing and GIS applications in municipal solid waste management
- Characterization, treatment and management of different types of solid waste: agricultural, municipal and e-waste
- Characterization of compost and the status of composting in India
- Presence of polycyclic aromatic hydrocarbons (PAHs) around informal e-waste recycling facilities and new methods for quantifying them
- An overview of the fundamentals of microbiology
- Mechanical-biological treatment in India and abroad.

There are many other issues pertaining to this vast and growing area of engineering research, development and management that remain to be addressed and we hope to include them in future volumes.

Kharagpur, India Sudha Goel

Preface to the First Edition

For the last few years, I have been struggling to put together a textbook on the basics of 'Solid and Hazardous Waste Management' thinking there was a niche waiting for it. It became increasingly apparent that there was sufficient published literature including textbooks that addressed basic concepts and fundamentals in this field. Yet, there remained gaps in the textbooks about major topics like the economics of waste management, management of specific waste streams like plastics, electronic waste, biomedical waste, standard analytical methods, mathematical and management methods in developing countries, etc. The list of topics not covered in textbooks was growing as colleagues, peers and students expressed their interests in working on different aspects of solid and hazardous waste management in the last 12–13 years. This book has 'happened' with several experts and authors from India and abroad contributing their efforts to the field and agreeing to share their work for this book on many of these diverse topics.

The objective of this book is to provide the reader a review of major current issues in solid and hazardous waste management that are not covered in standard textbooks. Several examples and case studies are provided to demonstrate the applicability of new concepts and ideas. In general, textbooks used by students are written by authors from developed countries. For students from developing countries, many of the concepts and ideas in these textbooks are 'alien' to them and difficult to assimilate and practice. Almost all examples in this book are based on experiences in India and it is hoped that good practices in developing countries will become apparent to readers.

Key features of the book include:

- The circular economy and the economics of solid waste management
- Remote sensing and GIS applications
- Characterization, treatment and management of different types of solid waste: institutional, biomedical, residential, e-waste, plastics and fly ash
- Characterization of compost
- An overview of the fundamentals of microbiology
- An overview of environmental regulations especially those pertaining to solid and hazardous waste management.

There are many other issues pertaining to this vast and growing area of engineering research, development and management that remain to be addressed. I will have to leave that to other experts and students in this field.

Kharagpur, India Sudha Goel

About This Book

Solid and hazardous waste management is a growing field in India and abroad and presents many technical, management and financial challenges in its implementation. This volume addresses the current status of municipal waste management in the country and the treatment methods used. Today the field encompasses many different waste streams like agricultural waste, biomedical waste, plastics, waste from thermal power plants and e-waste. Agricultural waste and the potential for deriving energy from it and e-waste management are two waste streams that are discussed in this book. The presence of polycyclic aromatic hydrocarbons (PAHs) in soils around informal e-waste recycling facilities is reviewed and advances in laboratory analytical methods for quantifying PAHs are covered in three different chapters. A new case study based on the use of remote sensing and GIS in municipal solid waste (MSW) management is also presented in this book along with a review of mechanical-biological treatment of MSW in India and abroad.

This book is written for students and practitioners in the field assuming that they are already familiar with the basics. The intent was to provide reviews, examples and case studies of innovative applications and methods in this growing area of research and development.

Key Topics Covered in the Book

- Remote sensing and GIS applications in municipal solid waste management
- Characterization, treatment and management of different types of solid waste: agricultural, municipal and e-waste
- Characterization of compost and the status of composting in India
- An overview of the fundamentals of microbiology
- Mechanical-biological treatment in India and abroad.

Contents

About the Editor

Sudha Goel is currently Professor in the Civil Engineering Department of the Indian Institute of Technology Kharagpur. She completed her B.E. (1990) in Environmental Engineering from L.D. College of Engineering, Ahmedabad and M.S.E. (1992) and Ph.D. (1996) in Environmental Engineering from The Johns Hopkins University in Baltimore, USA and has been working on various aspects of environmental engineering since then. She has worked on several consulting projects and taught various courses in India and abroad. Her research interests include solid and hazardous waste management, water quality and treatment, and environmental impact and risk assessment. She has several books and papers in national and international journals to her credit.

Chapter 1
Solid Waste Management: An Introduction

Sudha Goel

1 Introduction

Solid waste management is now acknowledged as one of the major environmental issues of our times. It remains a challenge for developed countries and is an exponentially growing problem for developing countries. The last four decades are marked by several incidents highlighting problems with solid and hazardous waste management across the world.

1. Transboundary shipments of hazardous and solid waste have received worldwide attention and media headlines. Infamous examples include the *Khian Sea* which started its journey from Philadelphia in 1986 and the *Mobro 4000* which started its journey from New York in 1987; two US barges or ships with cargoes of municipal solid waste that went from one country to another looking for a port to dump their wastes. The *Mobro* went all the way to Belize and brought its cargo back to Brooklyn, New York for incineration and landfilling.[1] Many speculate that the *Khian Sea* dumped its cargo somewhere in the Indian Ocean in 1988.[2]
2. A fire started on 27 January 2016 in India's oldest and largest open dumping site—Deonar in Mumbai and was visible in satellite images (shown in Fig. 1). It continued for several days leading to complaints of air pollution and closure of schools for 2 days. The massive fire was attributed to the buildup of methane gas in the open dump where the waste had reached heights of more than 30 m. Fires in this dumping site have been recurrent making living conditions in the surrounding areas extremely difficult.

[1] https://en.wikipedia.org/wiki/Mobro_4000.

[2] https://en.wikipedia.org/wiki/khian_sea_waste_disposal_incident.

S. Goel (✉)
Civil Engineering Department and School of Environmental Science and Engineering, Indian Institute of Technology Kharagpur, Kharagpur 721 302, India
e-mail: sudhagiitkgp@gmail.com

S. Goel (ed.), *Advances in Solid and Hazardous Waste Management*,
https://doi.org/10.1007/978-3-031-49144-3_1

Fig. 1 Plume from a massive fire in a solid waste dump in Deonar, Mumbai on 27 Jan 2016[3]

3. Fires have been recurring at two large solid waste dumpsites in Bhalswa and Ghazipur in Delhi.
4. Many sources have identified poor solid waste management as one of the major contributing factors to the spread of the plague in Surat, India in 1994 (see box for details). Incidents like these led to the promulgation of the Municipal Solid Waste (Management and Handling) Rules in India in 2000.

Yersinia pestis is the causative agent of the bubonic plague and is a parasite for the rat flea. The rat flea in turn is a parasite for rats which are generally attracted to food and other waste lying in the open. When a rat flea bites rats, humans or other mammals, it injects the bacteria into the organism leading to infection and death. The timeline of the plague in Surat is shown in the following table (Titball and Leary 1998).

[3] https://earthobservatory.nasa.gov/images/87429/fire-burns-in-mumbai-landfill.

Timeline of the plague in Surat (Titball and Leary 1998)

Timeline	Events
October 1993	Residents of Mamla village in Beed, Maharashtra left their homes due to fear of a Latur-type earthquake leaving piles of grain unattended. This may have attracted wild rats/rodents
August 1994	Flea nuisance and rat deaths were reported in Beed; soon after the first case of bubonic plague was reported in Beed. Similarly, an unusual number of rat deaths in Surat and soon after the first human death by pneumonic plague was reported
September 1994	Water logging due to poor SWM and possibly a person from Beed spread infection and Pneumonic plague in Surat city
October 1994	876 presumptive cases (596 in Maharashtra and 151 in Gujarat); no fatalities
	54 fatalities of which 52 in Gujarat, 1 in Delhi and 1 in Karnataka

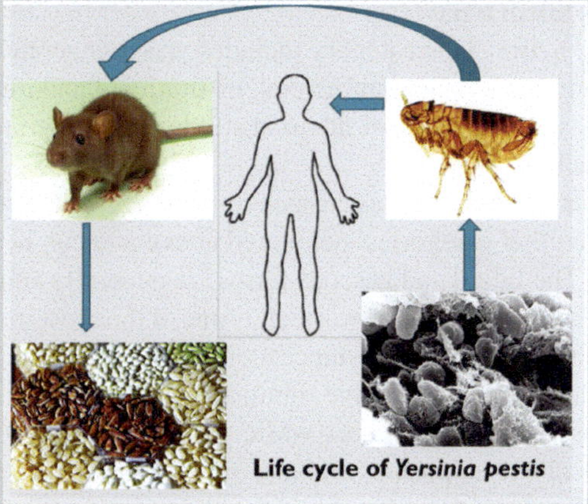

Life cycle of the bacterium, *Yersinia pestis*, the causative agent of bubonic plague (Photo of rat flea and Scanning Electron Micrograph of *Yersinia pestis*[4]).

5. Burning of crop stubble (mainly paddy residue remaining after harvest) in the northern states of Punjab and Haryana is common practice and has led to terrible air quality problems every winter in the entire Indo-Gangetic plain.

[4] https://en.wikipedia.org/wiki/yersinia_pestis.

1.1 What is Solid Waste?

Waste is defined as any material that is discarded by the possessor or generator due to its lack of value to them. Therefore, solid waste is defined as any waste material that is neither in a liquid nor gaseous state. Further, containerised liquid and gaseous wastes are also included in the term solid waste. Solid waste includes several categories of waste of which the most important ones are shown in Fig. 2 and include the following:

a. *Agricultural and allied industries*: Agrarian economies generate large amounts of agricultural waste during harvest and in food processing activities.
b. *Industrial and mining waste*: Several countries are heavily dependent on the mining of metals and allied industries for their economic growth. Mine tailings constitute a major portion of mining and industrial waste.
c. *Ash from thermal power plants*: Despite the growing share of renewable energy sources, coal-based thermal power generation remains the largest source of electricity worldwide. Burning of coal in thermal power plants generates large quantities of ash. Management and disposal of this ash is a major solid waste management problem today.
d. *Municipal solid waste*: This category includes waste generated in urban centers and derived from residential, commercial, institutional and municipal activities. It is a relatively small fraction of the total solid waste generated but remains its most visible fraction.
e. *Hazardous waste*: Hazardous waste is generated in comparatively small amounts compared to the other categories mentioned above and can be in solid, liquid or gaseous forms. The liquid and gaseous forms are generally containerised and are often considered a subset of solid waste. It is one of the most important categories of waste due to the severity of its impacts on public health and the environment. Any waste material that is reactive, ignitable, corrosive, toxic, radioactive, or infective is classified as hazardous waste. This category includes radioactive waste, biomedical waste, plastics and electronic waste (e-waste).

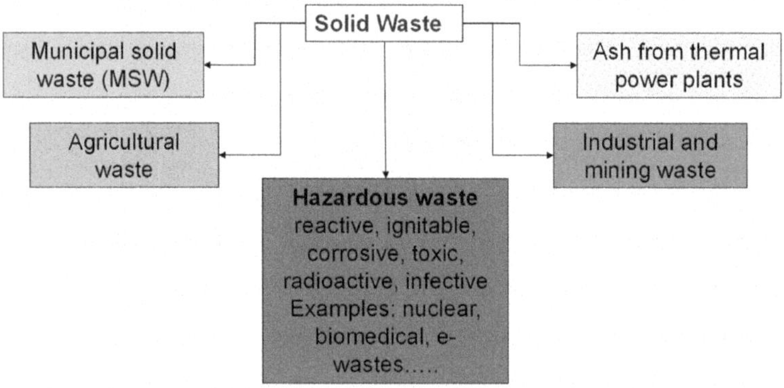

Fig. 2 Categories of solid waste

1.2 Importance of Solid Waste Management

Open dumping of municipal solid waste on land remains the norm rather than the exception in most low- and middle-income countries (Hoornweg and Bhada-Tata 2012; Kaza et al. 2018). Despite significant progress in building treatment facilities and landfills in the last couple of decades, open dumping remains the most common disposal option in India. In India, even after the Municipal Solid Waste (Management and Handling) Rules were promulgated in 2000 and replaced by the Solid Waste Management Rules, 2016, a large fraction of the total waste collected is dumped on open land. Problems associated with the open dumping of solid waste are enormous, long-term as well as short-term, and have far-reaching implications as evidenced by the aforementioned incidents. Poor solid waste management can lead to public health hazards, air, water and soil pollution, and a decrease in property values (Goel 2017).

2 Integrated Solid Waste Management

Integrated solid waste management takes a holistic approach that should be applied to all functional activities to be undertaken. Specifically, objectives that need to be addressed in integrated solid waste management include:

1. ***Minimise public health and environmental impacts***: by using appropriate collection, transport, storage and disposal technologies and practices.
2. ***Minimise resource consumption***: The concept of a circular economy is based on reducing resource consumption thereby reducing the amount of solid waste generated. This also includes the implementation of the 4Rs: Reduce, Reuse, Recycle, and Recover (Goel 2008). Another method for reducing resource consumption is adopting 'Green Design' principles for delivering goods and services (Masters and Ela 2008).
3. ***Adopt waste-to-energy technologies***: Since energy remains one of the most precious resources in today's world, any technology that allows the conversion of waste to useful energy is beneficial.
4. ***Minimise greenhouse gas emissions***: Another major issue today is climate change which is driven by the generation of greenhouse gases (GHG) like CO_2 and CH_4. Incineration of waste would result in the generation of GHG. If organic waste can be converted to end products like biogas or compost, there are net benefits due to a reduction in GHG emissions. Composting and biogas generation are also considered to be carbon–neutral technologies since there is no reliance on fossil fuels.
5. ***Minimise use of landfills and conserve land***: Land is a precious resource, and land use for landfilling waste must be avoided or minimised to the maximum extent possible. Landfilling has several adverse, long-term impacts on public health and the environment and requires long-term post-closure monitoring.

Minimising the use of landfills is a better option rather than creating landfills with more complex designs.

2.1 Principles

Integrated solid waste management includes several principles that can be incorporated into the design and creation of products and processes and in managing the wastes created during these processes.

2.1.1 Waste Management Hierarchy

The Waste Management Hierarchy defines the order of preference in which to use different waste management options. One of the objectives of Integrated Solid Waste Management is to reduce the amount of waste to be handled. Therefore, the SWM hierarchy shown in Fig. 3 has source reduction as the most preferred option followed by waste minimisation, reuse, recycling and waste-to-energy conversion. The least preferred options are treatment and disposal options like landfilling which do not result in useful end-products.

Source reduction is based on the concept that 'garbage that is not produced does not have to be collected'. Reduction in resource consumption is the main premise of source reduction and needs to be incorporated into the design of products and processes. Products/processes should be designed to minimise toxicity, weight (packaging), and energy use and should have a longer service life. The green

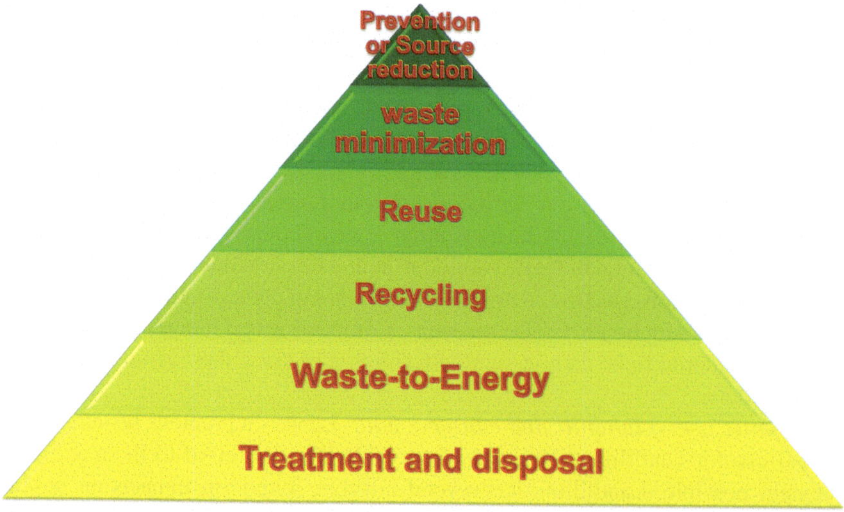

Fig. 3 Waste management hierarchy

design includes all of these concepts with the additional objective of minimising environmental impacts.

Waste minimisation requires that the amount of waste generated be reduced to the maximum extent possible. One example is segregation at source. By separating the waste into different fractions at the source, the amount of waste to be disposed off can be reduced or minimised. At a minimum, MSW should be segregated into three fractions: recyclables, biodegradable and non-biodegradable waste.

Reuse involves using the same product several times, perhaps for different uses. For example, containers and packaging materials like cardboard boxes and plastic bags and containers are often used multiple times before being discarded.

Recycling involves modification of the waste products to create newer products. Waste products made of plastic, paper, glass, and metals are easily recycled at the end of their lives.

Waste-to-energy is the conversion of waste materials into fuel or energy. The best examples are the use of solid waste for the generation of biogas or biodiesel which serve as fuels or the controlled burning of organic material in incinerators to generate heat and/or electricity.

Treatment and disposal is the least preferred option in the waste management hierarchy and is resorted to only in the absence of all other possibilities. The best example would be the use of controlled incineration followed by landfilling. If the waste material is low in moisture content and high in organic content, it can be incinerated under controlled conditions with appropriate air pollution control equipment. Ash and inert material remaining after incineration can then be landfilled.

2.1.2 The Four Rs

The four Rs of SWM, i.e., Reduce, Reuse, Recycle, and Recover are described here with some examples of current practices and some that can be introduced (Goel 2008). Some of these principles are part of the SWM hierarchy.

Reduce: The first principle is to reduce the amount of material required to deliver one unit of product without sacrificing its utility or quality. Examples include the use of refills in products like household consumables and the use of rechargeable batteries instead of disposable batteries. This principle also includes using materials that have a longer lifetime (reducing the use and need for disposables). Another method of implementing this principle would be to eliminate or reduce the need for certain products. Examples include substituting print media with electronic media and using gas pipelines instead of gas cylinders.

Reuse: The monetary incentives for implementing this principle are self-evident. Most people reuse containers of cardboard, plastic, metal, and glass for storage and other purposes in homes, offices, and business establishments. Packaging materials and furniture are similarly easy to reuse.

Recycle: Recycling requires separating different components of waste and reprocessing them into new products. Plastics, paper, steel and aluminum cans, glass bottles, and yard waste (composting) are examples of MSW components that can be recycled to create new products. Used furniture or wood is often re-cast into new products.

Recover: In general, recyclables should be separated at source. However, waste that is collected is unsegregated and desired materials have to be separated (recovered) at a central facility. In India, even though the bulk of recyclable materials are separated at source, it is becoming increasingly evident that some amounts of paper cartons, plastic, foil wrappers, and household hazardous items like batteries, tube lights, and thermometers are left in the mixed waste. A materials recovery facility where these materials can be separated from more benign materials like soil and organic matter is essential, especially if the waste is to be composted or biogas generated from it. Developed countries tend to have well-designed materials recovery facilities (MRFs).

2.1.3 Life Cycle Assessment (LCA)

Life cycle assessment also called 'cradle-to-grave' assessment is a method of accounting for all materials and energy inputs for any product during its entire life. This helps to identify outputs of solid waste, air and water pollutants and waste heat that can be recovered. Materials that are wasted in processing and subsequently, during product reuse and recycling can be recovered and the manufacturing processes can be made more efficient in terms of use of energy, air and water. Final disposal options for any product can be identified during a life-cycle assessment and a comparative analysis of products can be done using this tool to determine the best product in terms of its overall environmental impacts.

2.1.4 Extended Product Life

Products that have a longer service life lead to lower resource consumption and therefore, lower waste generation. The trend in the last few decades has, however, been towards the increasing use of disposables. Ink pens, which used to last for years were replaced by ballpens with refills that lasted for a few months, and these have now been replaced by ballpens or gel pens that are entirely disposable and cannot be refilled. Many other products are now manufactured with short service lives and manufacturers often discourage the use of refills for such products. Printer cartridges and disposable batteries can be easily refilled or recharged, respectively, and can reduce resource consumption as well as the price of the product. However, as the economy of a region/country improves, resource consumption is higher due to higher consumer spending levels and 'natural or planned obsolescence' by manufacturers, institutions, and businesses (Jena and Goel 2015). Most white goods manufacturers tend to produce goods with shorter service lives, leading to more waste generation. Examples of consumer goods designed for shorter service lives include computers, cell phones, clothing, electrical and electronic appliances (white goods) and motor vehicles. This natural tendency must be restrained by greener government policies and consumer willpower.

2.1.5 Material Selection and Durability

From an environmental standpoint, it is important to select materials and processes that will allow the components of the products to be reused or recycled easily. For example, it is possible to design car interiors with easy-to-disassemble recyclable plastics as is the case with BMW and Volkswagen in Germany (Masters 1998). Today, the best management practices for dealing with e-waste are to recycle plastics from computers and metals from printed circuit boards.

Minimising environmental impacts can entail substituting one material for another more environmentally benign option. This may at first be a more expensive proposition, but the long-term benefits can be enormous. A case in point is the use of steel, lead and copper telecom cables by AT&T in the 1950s (Masters 1998). Lead was replaced by polyethylene in the 1980s and fiber-optic cables replaced copper cables in the following decade, leading to large savings in lead, copper and energy use. Wireless technologies are another step in the same direction since they reduce resource consumption and eliminate destructive digging and defacing of roads, pavements and buildings. Today, cast iron pipes in water supplies are largely being replaced by plastic pipes due to their greater durability and fewer problems of corrosion and leakage. Likely, mercury-free thermometers and lights like light-emitting diode (LEDs) bulbs will be common products in the future since they are already available but are not cost-effective presently in comparison to fluorescent lights.

Improving material efficiency is another aspect of green design. An example of this is the improved efficiency of lead-acid car batteries, which provide the same performance with smaller amounts of lead. Similarly, reducing the weight of packaging or using new or different packaging materials is another method for reducing waste generation.

2.1.6 Process Management

Process management, also called environmental auditing, is a method of accounting for all inputs of raw materials and energy, and outputs of products and waste. Inefficiencies in the system (manufacturing processes) can be identified easily and the process made more efficient. This is somewhat similar to life-cycle assessment, where the same is done for a product rather than a process. Examples of improvements in process management are the replacement of old parts like motors with more efficient ones; plugging leaks, using more fuel-efficient systems; better materials management and use of sensor-based, high-efficiency lighting systems.

2.1.7 Distribution

Methods of distributing a product also determine the extent of waste generation. Packaging methods and transportation costs and their associated environmental impacts and energy requirements are major factors that can be improved to reduce waste generation. Transportation costs depend on the method of shipping (road, rail, sea or air), the type of product and packaging required, and the distance. A comparative cost analysis (direct and indirect or environmental costs) of all transportation options is necessary before choosing the best option. The use of recycled materials for packaging is another effective option for waste reduction. One of the simplest examples of reducing resource consumption and energy usage is the supply of domestic cooking gas by pipelines rather than gas cylinders.

2.1.8 Policy and Standardisation

Many policy initiatives have been taken in developed countries to reduce waste generation. Return policies for products like household appliances and automobiles and their packaging have been enforced in Germany while these have been adopted voluntarily by large corporations in the USA and India. Tax incentives are another option used by governments to make environment-friendly products comparable in terms of costs with similar products in the market. Consistent and standardised labeling and a credible authority for implementing these are other methods of promoting waste reduction. A recent example is the energy efficiency labelling system that is implemented in India by the Bureau of Energy Efficiency (BEE). BEE provides a star rating for appliances so that the consumer can make an informed choice that results in net energy savings and environmental benefits.

Standardised labelling is also essential for recycling different grades of plastic. Most plastic products have one of the symbols shown in Fig. 4 and have to be segregated accordingly for recycling.

In India, the Eco Mark Scheme of the Bureau of Indian Standards (BIS) was introduced in 1991 to label environmentally-friendly products. The certification is given on the basis of a cradle-to-grave assessment of the product based on the following criteria[5]:

- that they cause substantially less pollution than other comparable products in production, usage and disposal,
- that they are recycled, recyclable, made from recycled products or biodegradable, whereas comparable products are not,
- that they make significant contributions to saving non-renewable resources including non-renewable energy sources and natural resources compared with comparable products, and

[5] https://cpcb.nic.in/ecomark-criteria/.

- that the product must contribute to a reduction in adverse environmental impacts associated with the use of the product, and which are specifically set for each product category.

Which plastics are recyclable?

Our World in Data

Summary of plastic polymer groups, their common uses, properties and recyclability.
Numerical coding (from 1-7) is typically provided on plastic items and gives information of their polymer grouping below. Recyclability is based on common recycling schemes but can vary between countries as well as regionally within countries; check local recycling guidelines for further clarification.

Symbol	Polymer	Common Uses	Properties	Recyclable?
1 PETE	Polyethylene terephthalate	Plastic bottles (water, soft drinks, cooking oil)	Clear, strong and lightweight	Yes; widely recycled
2 HDPE	High-density polyethylene	Milk containers, cleaning agents, shampoo bottles, bleach bottles	Stiff and hardwearing; hard to breakdown in sunlight	Yes; widely recycled
3 PVC	Polyvinyl chloride	Plastic piping, vinyl flooring, cabling insulation, roof sheeting	Can be rigid or soft via plasticizers; used in construction, healthcare, electronics	Often not recyclable due to chemical properties; check local recycling
4 LDPE	Low-density polyethylene	Plastic bags, food wrapping (e.g. bread, fruit, vegetables)	Lightweight, low-cost, versatile; fails under mechanical and thermal stress	No; failure under stress makes it hard to recycle
5 PP	Polypropylene	Bottle lids, food tubs, furniture, houseware, medical, rope, automobile parts	Tough and resistant; effective barrier against water and chemicals	Often not recyclable; available in some locations; check local recycling
6 PS	Polystyrene	Food takeway containers, plastic cutlery, egg tray	Lightweight; structurally weak; easily dispersed	No; rarely recycled but check local recycling
7 OTHER	Other plastics (e.g. acrylic, polycarbonate, polyactic fibres)	Water cooler bottles, baby cups, fiberglass	Diverse in nature with various properties	No; diversity of materials risks contamination of recycling

Source: based on general US & UK guidelines, and chemical polymer properties. Icon graphics from Noun Project.
This is a visualization from OurWorldinData.org, where find data and research on how the world is changing.

Fig. 4 Different grades of plastic, their properties and recyclability

2.2 Functional Activities in Municipal Solid Waste Management

Figure 5 depicts the major functions or activities that are essential for municipal solid waste management (MSWM) and their relationship to each other. MSW Management is one of the most expensive essential municipal services provided by urban local bodies. On average, MSWM budgets in low-income countries constitute >19% of the total municipal budget, >11% in middle-income countries and >4% in high-income countries (Kaza et al. 2018).

2.2.1 Generation and Composition

Municipal solid waste is generated in urban areas and comes from several sources within the city such as residential, commercial, industrial, institutional and municipal sources. Sources of waste and their composition are described in Table 1.

Generation is the first activity in Municipal Solid Waste Management (Fig. 5) and MSW is the only waste stream where the generator of waste is not held responsible for the waste generated. The Urban Local Body (ULB) is responsible for the collection of all waste generated within the city or town limits and its segregation, transfer, treatment, and disposal. Generators of other categories of solid waste like biomedical, hazardous, batteries, and fly ash are held responsible for taking care of the waste generated and each of these waste streams is governed by its own set of rules.

In general, at the household level, waste that is generated is separated as recyclables and disposables—and is defined as 'separation at source'. Recyclables are stored and kept separately for selling to kabadiwalas (scrap dealers in India), or

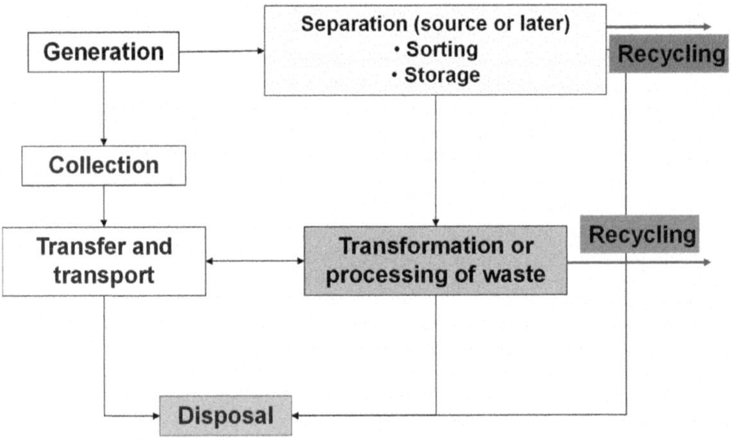

Fig. 5 Activities in municipal solid waste management systems

Table 1 Sources of municipal solid waste in urban areas (IBRD 1999)

Source	Typical waste generators	Types of solid wastes
Residential	Single and multiple family dwellings	Food, paper, cardboard, plastics, textiles, leather, yard wastes, wood, glass, metals, ash, special waste (e.g., bulky items, consumer electronics, white goods, batteries, oil, tires), and household hazardous waste)
Industrial and manufacturing	Light and heavy manufacturing, fabrication, construction sites, power and chemical plants	Housekeeping waste, packaging, food, construction and demolition materials, hazardous waste, ash, special waste
Commercial	Stores, hotels, restaurants, markets, office buildings	Paper, cardboard, plastics, wood, food wastes, glass, metals, special waste, hazardous waste
Institutional	Schools, hospitals, prisons, government centers	Same as commercial
Construction and demolition	New construction sites, road repair, renovation sites, demolition of buildings	Wood, steel, concrete, dirt
Municipal services	Street cleaning, landscaping, parks, beaches, other recreational areas, water and wastewater treatment plants	Street sweeping, landscape and tree trimmings, general wastes from parks, beaches, and other recreational areas, sludge
Agriculture	Crops, orchards, vineyards, dairies, feedlots, farms	Spoiled food wastes, agricultural wastes, hazardous wastes (e.g., pesticides)

recycling businesses. Further, segregation at source into two fractions of disposables: biodegradable and non-biodegradable, is now mandatory in all cities in India as part of the Solid Waste Management (SWM) Rules, 2016.

The average per capita mixed waste generation rate at the residential or household-level was 0.3 kg/cap-d in a survey of 21 households in Autumn 2020. City-level per capita generation rates are much higher than MSW generation rates at the household level as they include residential waste along with waste from public activities like fairs and festivals, industrial and commercial activities, street litter and sweepings, waste from public spaces like gardens, parks, stadia, treatment plants, and construction and demolition waste.

Waste Generation Rates

The total amount of MSW generated globally was 2.0 billion tons/y and approximately 1/3 was not managed appropriately (Kaza et al. 2018). High-income countries have only 16% of the world population but they generate about 34% of the total global MSW. Per capita waste generation ranged from 0.11 kg/cap-d in Sub-Saharan Africa to 4.54 kg/cap-d in North America while average generation rates ranged from 0.46 to 2.2 kg/cap-d with a global average of 0.74 kg/cap-d. Per capita waste generation rates are known to be correlated with income levels, both at the individual and country or regional levels (Kaza et al. 2018; Goel et al. 2017; IBRD 1999). The top three countries in terms of MSW generation were Bermuda, Canada and USA.

Per capita MSW generation rates in India were estimated to vary between 0.2 and 0.87 kg/cap-d in 2011 with an average generation rate of 0.5 kg/cap-d which has now increased to 0.6 kg/cap-d. Most of the MSW generated (72.5%) in India comes from Class I or larger cities.

The success of many treatment methods noted later in this section is dependent on the separation of waste at source into 'biodegradable' and 'non-biodegradable' or 'other or mixed' fractions (Hazra et al. 2013). Separation into biodegradable and other fractions is rarely practiced in India and treatment methods like composting, incineration and biogas generation whose success depends on this type of source separation tend to fail in India. In contrast, the separation of biodegradable and non-biodegradable materials is common in developed nations accounting for the success of their centralised treatment systems.

Estimate total MSW generation rates

To estimate total or per capita MSW generation rates for the future, it is necessary to forecast the population data for an area. MSW is an urban phenomenon and therefore, the population growth rate for urban areas is used.

Problem 1

Census of India data for total and urban populations in 2001 and 2011 are shown in the following table. Using the exponential equation: $N = N_0 * \exp(kt)$, determine the exponential growth rate for the urban population. Based on this growth rate, estimate the urban populations in 2021 and 2051.

Solution

The annual average exponential growth rate for the urban population based on the populations in 2001 and 2011 was found to be 2.79%. Based on this growth rate, the urban population in 2021 was 498 million and in 2051, it is estimated to be 1.15 billion. The urban population in India is expected to be close to 50% of the total population of the country in 2051.

Year	Total population	Urban population	Urban population as a % of the total population
2001	$1.03 * 10^9$	$2.85 * 10^8$	27.78
2011	$1.21 * 10^9$	$3.77 * 10^8$	31.16
2021	$1.43 * 10^9$	$4.98 * 10^8$	34.95
2051	$2.33 * 10^9$	$1.15 * 10^9$	49.32

Problem 2

Estimate the amount of MSW generated in India in 2021 and 2051.

Solution

The total MSW generation rate in urban areas can be determined by multiplying the population by the average per capita MSW generation rate. Therefore, assuming an average per capita MSW generation rate of 0.6 kg/capita-d, the total MSW generated in India's urban areas in 2021 would be 300 million kg/d and in 2051, it is estimated to be 690 million tons/d.

Problem 3

Determine the per capita waste generation rate for a city with a population of 5 million and a total MSW collection rate of 2500 tons/day.

Solution

In many cities, the per capita MSW generation rate is not known and cannot be quantified but the total amount of waste collected is known. We can assume that the waste collected is equal to the waste generated. Therefore, for a city where 2500 tons/day of MSW are collected and has a population of 5 million, one can determine the average per capita waste generation rate. In this case, the average per capita MSW generation rate is 0.5 kg/cap-d.

MSW Composition: Physical

The composition of MSW will determine the design and nature of transport vehicles, and choice of treatment and disposal methods. Physical composition of MSW is determined by taking a bulk sample of 25 kg or more and segregating individual recyclable fractions manually. Each fraction is weighed and its percentage of the total MSW is calculated. In a survey of nine households in 2020, the physical composition of residential waste was determined over a period of 10 days and the results are shown in Table 2.

Characterization of MSW

Other parameters that are included in the composition of MSW include moisture content, elemental analysis, calorific value, and proximate analysis which includes total solids, volatile and fixed solids. Methods for determining each of these parameters are described here.

Moisture Content: Samples of 100 g were taken in triplicate, and dried to a constant weight in an oven at 105 °C for 24 h, cooled in a desiccator and the difference in weight recorded (Vesilind et al. 2003). Moisture content is the percent sample weight lost in drying.

Elemental analysis: A CHNS-O Elemental analyser (Euro Vector EA3000) was used to determine the elemental composition of MSW samples. For Carbon, Hydrogen, Nitrogen and Sulfur determination, dried MSW samples were weighed and added to tin capsules before dropping them in the combustion chamber of the

Table 2 Average physical composition of household MSW in 2020

Waste fraction	Average for 9 locations and 10 days (n = 90), grams/d	Standard deviation, grams/d	Composition as a % of the total waste
Paper	7.94	7.40	2.41
Plastic	8.82	12.19	2.68
Cardboard	16.21	22.09	4.92
Textiles/clothes	27.34	80.38	8.29
Rubber and leather	40.02	116.27	12.14
Glass	21.01	60.10	6.37
Metals	6.74	16.33	2.04
E-waste	0.55	0.85	0.17
Miscellaneous	15.25	15.80	4.62
Food waste	133.24	108.04	40.40
Garden waste	52.65	81.87	15.97
Total	329.77	396.50	100.00

instrument. A thermal conductivity detector (TCD) is attached to the instrument to detect the different elements.

Calorific Value: Calorific value was determined in the laboratory using a bomb calorimeter which provides the gross calorific value at constant volume (Worrell and Vesilind 2012).

Proximate analysis (Total Solids, Volatile Solids, Fixed Solids): The same methods recommended for river and lake sediment samples and sludge samples, i.e., solid and semi-solid samples from water and wastewater treatment were used (Method 2540 G, APHA et al. 2005). Total solids are defined as the solids left in the sample after it has been dried to a constant weight at 103–105 °C. When total solids are ignited in a muffle furnace at 550 ± 50 °C, the fraction of solids remaining is defined as fixed solids and the fraction lost during ignition is termed volatile solids.

MSW samples were collected in Kharagpur from 4 Wards on 5 working days in 2019 following these methods and the results are shown in Fig. 6.

Problem 4

For designing MSW treatment and disposal facilities, it is essential to know the composition of the waste. The following table provides data regarding the composition of the waste and moisture content and energy content of the different fractions of MSW. These data can be used to determine the moisture content and energy content of the mixed waste.

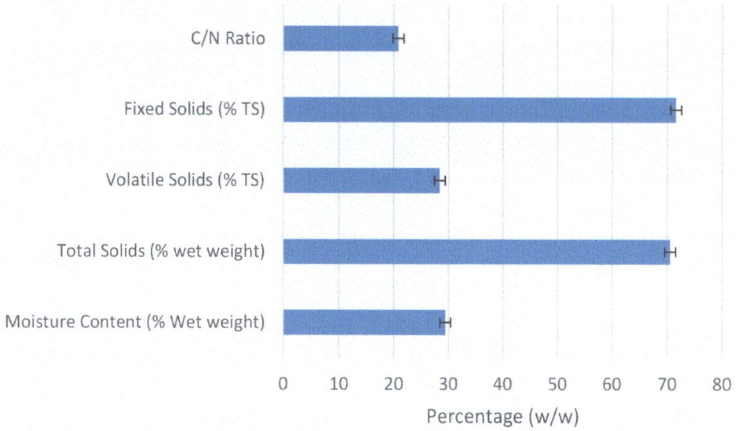

Fig. 6 Average MSW composition ($n = 20$) in Kharagpur in 2019

Waste fraction	% of fraction by wet weight	Moisture content, %	Energy content, kJ/kg
Paper and cardboard	15.00	7.00	16300
Plastics	0.75	2.00	0
Ferrous metals	0.80	3.00	0
Glass	0.40	2.00	0
Sand and fine earth	35.00	8.00	0
Compostable matter	37.5	60.00	5800.00
Others	10.55	5.00	0
Total	100.00		

Solution

For the data shown in the table, the moisture content of the mixed waste is 27% and the energy content is 4620 kJ/kg.

2.2.2 Segregation and Recycling

Waste collected can be segregated either at the source or at any later stage of collection or transportation. In general, households and small businesses in India segregate (or separate) waste at source. All waste materials that are recyclable/saleable like paper, plastic, metal and glass are sold by householders and business people to recyclists (or recyclers) or kabadiwalas, who move from door-to-door buying these waste materials. A large, informal but highly organised network of recyclists and businesses has evolved over the years to make this a highly efficient and financially viable method of recycling, especially in developing countries. However, as standards of living increase, these financial incentives become less significant or in many cases are no longer financially viable. This leads to the conditions in developed countries where recycling in residential areas remains a voluntary exercise with no financial incentives.

In the Indian cities, small amounts of recyclable materials that remain in the waste are collected by ragpickers at open dumps while those that are taken to the municipal treatment and processing facilities are segregated manually.

Segregation at source or at a materials recovery facility is generally done in developed countries, especially for biodegradable and non-biodegradable materials.

Recycling of various materials is constrained by several factors

1. Economics

   a. *Financial sustainability*: Sufficient income has to be generated in the recycling of various waste materials so that it remains a sustainable source of livelihood for those involved.
   b. *Quality of recyclable materials*: Another economic issue is the quality of the materials that are to be recycled. For example, newsprint is easier to recycle, i.e., gives better monetary returns, compared to magazine paper. Newsprint is manufactured by physical pulping of woody materials and no chemicals are added in the process. Magazine paper, on the other hand, goes through various physical and chemical processes where chemical agents like bleach are added to make it white and clay additives are added to make it glossy after physical pulping. The percentage of reusable pulp per unit weight of magazine paper is therefore much less compared to newsprint. In other words, after use, the pulp available from newsprint is much higher in quality and quantity compared to that obtained from magazine paper.
   c. *Quantity of recyclable materials*: The quantities of recyclable materials have to be sufficient to ensure long-term financial viability. The best example of this issue is the refusal of recyclists in India to pick up polyethylene terephthalate (PET) bottles but not high-density polyethylene (HDPE) bottles. The weight-to-volume ratio (specific weight) of PET is much less than that of HDPE. Therefore, the uncrushed PET bottles occupy more space (or volume) than they are worth in terms of the weight of the material, making it financially unattractive to transport them to recycling centers/businesses.

2. Ability to meet specifications
   A major issue in the recycling industry is whether the recycled product can meet specifications and compete with products made from virgin materials. For example, recycled plastic cannot be used for food packaging due to the presence of pigments and other additives in it. Similarly, recycled paper is generally restricted in terms of uses since it cannot be used for printing, and other high-end purposes. It is generally used as art paper, for blotting purposes, wrapping paper, paper bags and other low-end uses.
3. Convenience to sellers and buyers
   The recycling industry has a stronger presence in less developed countries compared to developed countries since there is a firmly established network of

recyclists and higher-order businesses. The door-to-door collection of recyclables with financial incentives for sellers and buyers in India makes it extremely convenient for people to recycle waste materials. Easily accessible deposit boxes for used batteries, cell phones, light plastics, computers, printers, and other disposable white goods help to reduce the burden of transporting these recyclables. Collection of large volumes or weights of these recyclable materials can also ease their recyclability in terms of quality, quantity and cost-effectiveness.

2.2.3 Collection

Disposable waste materials are collected and stored at home for pick-up by the municipal (or administrative) authorities or are deposited in large community vats, containers or bins. Collection efficiencies in India are extremely poor and range from 70 to 90% in major metropolitan cities and below 50% in smaller cities (Department of Economic Affairs 2009). However, this has improved significantly in the last 2 decades and based on annual CPCB reports, MSW collection efficiencies in 2019 were >98% in India. At the global level, high-income countries have high collection efficiencies (>90%) and spend less than 10% of their MSWM budgets on collection. Most of their MSWM budgets are spent on processing and disposal (Hoornweg and Bhada-Tata, 2012). In comparison, low- and middle-income countries have lower collection efficiencies and much higher percent spending on MSW collection ranging from 50 to 90% of their total MSWM budgets.

Waste can be collected from individual households by house-to-house collection, or the residents may be required to bring their waste to large community bins from where the ULB arranges to pick it up. Both systems are followed in both developed and developing countries. The first method is possible where residential units are single household units while the latter is followed in high population density areas, especially large apartment buildings. Collection is generally weekly or at higher frequency (daily is desirable) and storage of waste becomes an essential part of the collection process. The design of collection bins, their storage capacities, and their locations are dependent on the frequency of cleaning and the population to be served (Goel 2008). Separate collection systems are required for wastes from large commercial, industrial and construction activities.

MSW in cities is collected by the respective municipalities and transported to designated disposal sites, which are frequently low-lying areas on the outskirts of the city without taking any precautions or operational controls. Most of the MSW generated in India is directly disposed off on land in an unsatisfactory manner. Such inadequate disposal practices lead to problems that will impair human and animal health and result in economic and environmental losses. Such examples can be seen in almost every city in the country. The limited budgets earmarked for the municipalities make them ill-equipped to provide for the high costs involved in the collection, storage, treatment, and proper disposal of MSW. As a result, a substantial part of the MSW generated remains unattended at poorly maintained collection centres. The choice of a disposal site is more a matter of what is available rather than what is suitable.

2.2.4 Transfer and Transportation

Waste collected from containers or home bins is then transferred and transported to central facilities where appropriate processing or treatment methods can be applied. In developed countries, transfer stations are often combined with materials recovery facilities where different waste components are separated and sold for recycling.

Large urban areas may require transfer stations to improve collection and cost efficiency. Waste from community bins and individual homes is generally collected by small vehicles and deposited at transfer stations. From transfer stations, mixed waste can be segregated and taken either for further processing or for direct disposal to landfills.

The separated materials may include recyclables like paper, plastic, and metals which can be processed and transformed for the manufacture of new products and these require separate facilities for that. Composting of the organic fraction of MSW is often done either at the transfer stations or at the disposal sites.

2.2.5 Processing or Treatment of Waste

Several methods for treating MSW are available and are summarised in Fig. 7. MSW in India, especially if it is unsegregated, tends to be low in organic content and highly variable in moisture content depending on weather conditions (Kumar et al. 2009). For example, samples collected from open containers in the monsoon have very high moisture content but those collected in other seasons may not have high moisture content.

For wastes with low organic and low moisture content, the only option is landfilling. This applies to waste that is predominantly inert material like soil and dirt. For wastes with high organic and moisture content, biogas or biofuel generation is the best treatment strategy since these two factors are necessary for biological processing,

	Low Moisture Content	High Moisture Content
Low Organic Content	Landfilling	Composting
High Organic Content	Incineration/ Combustion	Biogas or biofuels

Fig. 7 Possible treatment options for different types of solid wastes

especially under anoxic (anaerobic) conditions. Wastes with relatively low or medium organic content but high moisture content are best composted while wastes with low moisture content but high organic content can be incinerated or combusted under controlled conditions with adequate air pollution control equipment.

Waste-to-Wealth (WTW) Conversion

A major global trend is the use of WTW methods for MSWM. WTW methods include processes for generating energy (Waste-to-Energy, WTE) or useful end products like compost. In a time of growing energy needs and high costs of energy, WTE represents an additional renewable energy source and one that is carbon–neutral or results in the reduction of GHG emissions. Waste can be converted to energy either chemically or biologically by several methods as shown in Fig. 8.

Chemical methods for WTE include combustion or incineration, gasification and pyrolysis where the objectives in MSWM are reduction in weight or volume of waste generated and generation of energy or other conversion products (Tchobanoglous et al. 1993).

Incineration is done by providing excess oxygen or air (based on stoichiometric requirements), gasification is by providing sub-stoichiometric quantities of air or oxygen and pyrolysis is done in the absence of air. Incineration can reduce MSW volume by 80–90% depending on the nature of the source material and the energy generated can be utilised in the form of electricity or steam. It is a waste-to-energy option that has been used successfully in many countries for municipal solid waste

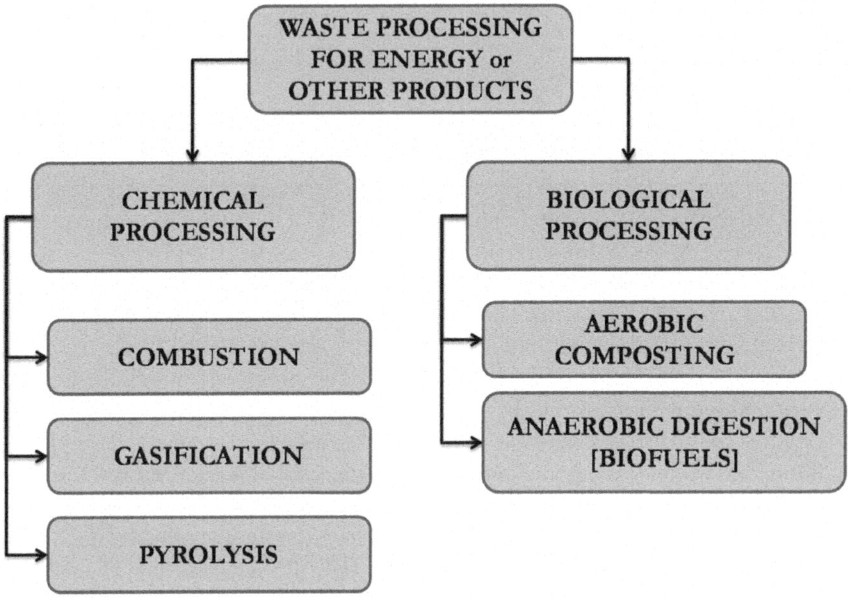

Fig. 8 Waste-to-wealth (WTW) conversion processes

management and is becoming increasingly popular. However, long-term success under Indian conditions has not been proved, mainly due to the need for and absence of air pollution control equipment. The minimum calorific value recommended for combustion-based WTE is 1790 cal/g. However, mixed MSW in India generally has a low calorific value (average of 1751 cal/g) and high moisture content (average 47%) which makes combustion or incineration a difficult proposition. On the other hand, in European countries where land is becoming scarce, this is becoming an increasingly popular treatment strategy (Annepu 2012). The costs of incineration are estimated to be $\geq$3 times more than landfilling, making landfilling the most preferred option even in developed countries where land area is still available (Hoornweg and Bhada-Tata 2012). However, in rich Western European countries where land prices are high, incineration is now the preferred option as it helps to meet the current global goal of 'Zero-waste to landfills'.

Gasification results in the production of a usable fuel—synthetic gas (or syn gas) which is a combination of H_2 and CO with small amounts of methane and carbon dioxide. Pyrolysis can be used to generate pyrolytic oil or gas with similar composition as syn gas. Recent estimates show that the energy recovery potential from MSW alone in India is 1460 MW while the current installed capacity is only 24 MW (Kalyani and Pandey 2014). Waste-to-energy methods like biogas generation apply only to agricultural and food processing waste and have not had commercial success with mixed municipal solid waste mainly due to lack of separation at source. It is estimated that one cubic meter of biogas can be converted to 1.5 kW of electricity (Annepu 2012).

2.2.6 Disposal

After most of the waste has been treated and compost or energy has been derived, the remaining material has to be landfilled. The current global goal is 'Zero-waste to landfills' and several European countries such as Germany, Switzerland, Belgium, Sweden and the Netherlands are sending less than 2% of their total MSW to landfills (Bufoni 2017). A landfilling target of less than 10% of the total MSW has to be achieved in Europe by 2035. The Netherlands has adopted a treatment and disposal strategy of recycling, composting and waste-to-energy conversion so that the amount of MSW that is sent for landfilling is nearly zero (Annepu 2012). Waste materials can be reused, recycled, incinerated or landfilled. The latter two options are the final disposal options with ash from the incineration process also requiring landfilling. In countries like Australia, Canada and the USA where land is available in plenty, landfilling has been the method of choice for the disposal of waste (Masters and Ella 2008). However, greater understanding and cost of landfilling has now made countries like the USA move away from this option resulting in a significant decline in the number of landfills over the last few decades (USEPA 2014, 2005). The number of landfills in 1988 in the USA was 7924 and dropped to less than 1200 in 2018. Ten

European countries were sending less than 1% of their total MSW in 2019 to landfills. These countries are Germany, Switzerland, Belgium, Sweden, the Netherlands, Norway, Denmark, Finland, Luxembourg and Austria.

Indian regulations require that only non-biodegradable materials should be landfilled. Further, biodegradable materials should be separated at the source and diverted for composting, anaerobic digestion or incineration. The market share of different MSWM technologies in India was estimated to be composting (50%), anaerobic digestion (30%), pelletisation (10%) and sanitary landfilling (10%) (Singh et al. 2011). Financial resources for better MSWM are necessary and current budgetary allocations are obviously insufficient for achieving high performance. Currently, municipalities are dependent on property and other taxes for their revenues which are generally insufficient for providing the required level of municipal services. There is sufficient evidence that people are willing to pay more for better services (Hazra et al. 2013).

As part of the Solid Waste Management Rules, 2016, there are 10 service level benchmarks (SLB) that are now defined for all ULBs to measure their performance. A list of these service-level benchmarks is provided here:

1. **Waste minimisation and reduction at the source**.
2. **Door to door collection** of waste from all residential, commercial, and institutional establishments in **covered vehicles**.
3. **100% Segregation** of waste at source.
4. **Covered Bins** for secondary collection.
5. **Secondary collection and transportation** in covered vehicles.
6. **Covered waste transfer/storage stations**.
7. **Material recovery/recycling** in accordance with SLBs for MSW.
8. **Scientific processing** in compliance with SWM Rules 2016.
9. **Scientific disposal** of inert material in compliance with SWM Rules 2016.
10. **100% cost recovery of Operation & Maintenance (O&M)** expenses from user charges as mandatory urban reform stipulates and **90% collection efficiency of user charges** in accordance with SLB for MSW.

2.2.7 The Cluster Approach

Municipal solid waste treatment and disposal facilities have been set up in most large metropolitan centers in India in the last few decades. The emphasis in setting up these facilities has been to create technically and financially self-sustaining facilities which has been a major challenge for ULBs all over the country. Several ULBs in the country including those in large metropolitan cities were not able to abide by the SWM Rules, 2016 due to lack of trained manpower, lack of financial resources, inadequate waste quantities for sustainable long-term treatment, and lack of segregation at source.

Therefore, a new approach towards sustainability has been adopted called the 'Cluster Approach' where consortia of 3 or more ULBs, generally from small towns and cities, are formed to install and operate MSW treatment and disposal facilities. These consortia have been formed successfully in Uttarpara-Baidyabati (close

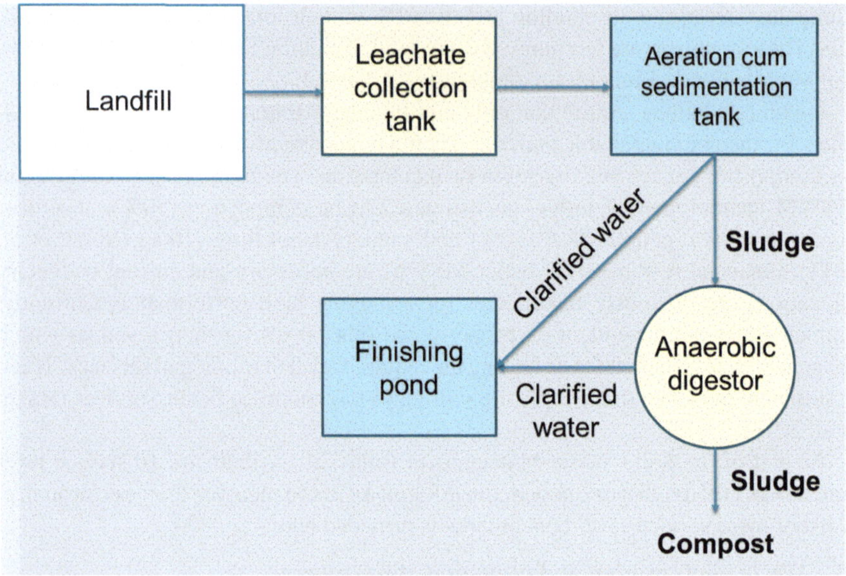

Fig. 9 Flowchart for the landfill and leachate treatment facility at Baidyabati

to Kolkata; see Chap. 4, Fig. 8) and in Madhya Pradesh, Gujarat, Rajasthan and Uttarakhand. Most of these facilities are in the development phase while the one in the Kolkata Metropolitan Area has been completely operational since 2017.

The Kolkata Metropolitan Area is managed by the Kolkata Metropolitan Development Authority (KMDA) and includes most of the suburbs of Kolkata City. KMDA along with 6 municipalities in Hooghly District, namely Champdani, Baidyabati, Serampore, Konnagar, Rishra and Uttarpara-Kotrung formed a consortium for managing MSW. With assistance from the Japan International Cooperation Agency (JICA), a covered windrow composting plant was setup in Uttarpara and a sanitary landfill and leachate treatment facility was installed in Baidyabati and became functional in 2017. A flowchart of the units in the Baidyabati landfill site is shown in Fig. 9.

3 Recommendations for MSW Management

With India's urban population increasing at an average annual rate of 3%, total solid waste generation rates are going to increase at a much greater rate with an economy growing at the rate of 5–10% in the last 3–4 decades. Source reduction and implementation of the 4Rs is an essential part of integrated solid waste management and needs all stakeholders—residents, businesses and administration to work together towards it.

The success of technologies like composting, anaerobic digestion and incineration depends on source separation into 'biodegradable' or 'organic' and other fractions. This will require public awareness campaigns to educate people and businesses about the need for doing so.

Given the high population density, increasing urban population and MSW generation, and the current goal of 'Zero-waste to landfill', sanitary landfilling is going to become obsolete or a prohibitively expensive proposition for the country in the future. The untapped potential in WTW needs to be exploited to the fullest extent, which means that investment in terms of financial and technical resources for developing waste-to-energy systems is essential for the future.

For creating technical and financial self-sustaining systems, the Cluster Approach has been successful in the KMDA area and can be emulated in other parts of the country.

Concluding Remarks

In summary, it is clear that solid waste generation is increasing exponentially all over the world due to increasing population and higher per capita resource consumption. While developed countries have sustainable infrastructure in place for handling these types of wastes, lesser developed countries have a long way to go. The search for waste-to-wealth treatment technologies and carbon–neutral processes and solutions remains a global challenge and given the current goal of 'Zero-waste to landfill', this is more important than ever before.

References

Annepu, R.K. (2012). Sustainable Solid Waste Management in India. MS Thesis, Department of Earth and Environmental Engineering. Columbia University.

APHA, AWWA, and WEF (2005). Standard Methods for the examination of water and wastewater.

Bufoni, A.L. (2017). The change from aerobic to anaerobic landfilling and increase of correspondent emissions and expenses. *In*: Sengupta, D. and S. Agrahari (eds.). Modelling trends in solid and hazardous waste management, Edition: 1st; Chapter: 8. Springer-Nature.

Department of Economic Affairs (2009). Position Paper on the Solid Waste Management Sector in India. Ministry of Finance, Government of India.

Goel, S. (2008). Municipal solid waste management (MSWM) in India: A critical review. *Journal of Environmental Science and Engineering, NEERI*, **50**(**4**): 319–328.

Goel, S. (2017). Solid and hazardous waste management: An introduction. *In:* Advances in solid and hazardous waste management, Springer Nature.

Goel, S., Ranjan, V.P., Bardhan, B. and T. Hazra (2017). "Forecasting Solid Waste Generation Rates." *In*: Modelling Trends in Solid and Hazardous Waste Management, pp. 35–64. Springer Singapore. https://doi.org/10.1007/978-981-10-2410-8_3.

Hazra, T., Goel, S. and B. Maitra (2013). Willingness-to-pay for solid waste management service attributes: Kolkata Municipal Corporation Area, India, as a case study. *International Journal of Environment and Waste Management*. **12**(**4**): 406–421. https://doi.org/10.1504/IJEWM.2013.056627.

Hoornweg, D. and P. Bhada-Tata (2012). What a waste: A global review of solid waste management. *Urban Development Series Knowledge Papers*. Washington DC, USA. https://doi.org/10.1111/febs.13058.

IBRD (1999). What a waste: Solid waste management in Asia. World Bank. Washington, DC.

Jena, S.K. and S. Goel (2015). E-waste generation in an academic campus: IIT Kharagpur as a case study. *Pollution Research,* **34**(**2**): 315–320.

Kalyani, K.A. and K.K. Pandey (2014). Waste to energy status in India: A short review. *Renewable and Sustainable Energy Reviews*, **31**: 113–120. https://doi.org/10.1016/j.rser.2013.11.020.

Kaza, S., Yao, L., Bhada-Tata, P. and F. Van Woerden (2018). What a waste 2.0: A global snapshot of solid waste management to 2050." Washington DC. https://doi.org/10.1596/978-1-4648-1329-0.

Kumar, S., Bhattacharyya, J.K., Vaidya, A.N., Chakrabarti, T., Devotta, S. and A.B. Akolkar (2009). Assessment of the status of municipal solid waste management in metro cities, state capitals, class I cities, and class II towns in India: An insight. *Waste Management*, **29**(**2**): 883–895. https://doi.org/10.1016/j.wasman.2008.04.011.

Masters, G.M. (1998). Introduction to Environmental Engineering and Science. NJ, US: Prentice Hall India.

Masters, G and W Ella (2008) Introduction to Environmental Engineering and Science, Prentice Hall India.

Singh, R.P., Singh, P., Araujo, A.S.F., Ibrahim, M. and O. Sulaiman (2011). Management of urban solid waste: Vermicomposting a sustainable option. *Resources, Conservation and Recycling*, **55**(**7**): 719–729. https://doi.org/10.1016/j.resconrec.2011.02.005.

Tchobanoglous, G, H Theisen, and S Vigil. 1993. Integrated Solid Waste Management: Engineering Principles and Management Issues. Singapore: McGraw Hill International Ed. Civil Eng Series Singapore.

USEPA (2005). Municipal Solid Waste Generation, Recycling, and Disposal in the United States: Facts and Figures for 2003. Washington DC, USA. www.epa.gov/osw.

USEPA. 2014. "Municipal Solid Waste Generation, Recycling, and Disposal in the United States: Facts and Figures for 2012." Washington DC, USA. Available from www.epa.gov/osw.

Titball, R.W. and S.E.C. Leary (1998). Plague. *British Medical Bulletin*, **54**(**3**): 625–633.

Chapter 2
Electronic Waste (E-Waste) Generation and Management

Deblina Dutta and Sudha Goel

1 Introduction

Rapid advancement in technology, especially the production of electrical and electronic equipment has resulted in a new stream of waste known as electrical and electronic waste (E-waste) making it the fastest-growing waste stream in recent times. E-waste, if managed improperly, causes adverse effects on the global environment as well as on human health.

Discarded E-waste was classified into 6 categories and the global distribution of this waste is shown in Table 1 (Forti et al. 2020a, b):

i. Temperature exchange equipment which includes refrigerators, freezers and air conditioners, and heat pumps;
ii. Screens, monitors, and equipment containing screens such as televisions, monitors, laptops, and tablets;
iii. Lamps such as fluorescent lamps and LED lamps;
iv. Large equipment such as washing machines and dryers, dishwashing machines, large printing devices, and photovoltaic panels;
v. Small equipment which may include vacuum cleaners, microwaves, ventilation equipment, toasters, electric kettles, electric shavers, scales, electric toys, small medical devices, and control instruments; and
vi. Small IT and telecommunication equipment such as mobile phones, personal computers, printers, game consoles, calculators, and other small IT equipment.

D. Dutta
Department of Environmental Science and Engineering, School of Engineering and Sciences, SRM University-AP, Amaravati, Andhra Pradesh 522 240, India

S. Goel (✉)
Civil Engineering Department and School of Environmental Science and Engineering, Indian Institute of Technology Kharagpur, Kharagpur 721 302, India
e-mail: sudhagiitkgp@gmail.com

S. Goel (ed.), *Advances in Solid and Hazardous Waste Management*,
https://doi.org/10.1007/978-3-031-49144-3_2

Table 1 Categories of E-waste and quantities generated world-wide (Forti et al. 2020a, b)

Category	Description	Quantity, Million metric tons (MT)	Percentage of total, %
1	Temperature exchange equipment	10.80	20.15
2	Screen and monitors	6.70	12.50
3	Lamps	0.90	1.68
4	Large equipment	13.10	24.44
5	Small equipment	17.40	32.46
6	Small IT and telecom equipment	4.70	8.77
		53.60	100.00

Typical weights and lifespans of some of the most common electronic waste items are summarized in Table 2. E-waste covers more than 1000 materials which can be categorized as 'hazardous' or 'non-hazardous'. Mostly, it consists of ferrous and non-ferrous metals, plastics, glass, plywood, printed circuit boards, rubber, concrete, and ceramics, etc. Iron and steel constitute a major part of Waste Electrical and Electronic Equipment (WEEE).

2 Problems Associated with E-Waste

Proper care, management and handling procedures should be considered when dealing with E-waste. Improper handling may cause serious damage to human health as well as to the environment and several reviews on the topic are available (Maphosa and Maphosa 2022; Singh et al. 2022). Table 3 shows some of the constituents present in E-waste and their effects on human health and the environment.

3 E-Waste Generation

3.1 Global Perspective of E-Waste Generation

According to the Global E-waste Monitor, the generation of E-waste was estimated to be 53.6 million metric tons (Mt) in 2019 at the global level with a per capita generation rate of 7.3 kg/cap-y (Forti et al. 2020a, b). This quantity is expected to double in 16 years. Asia was the largest producer of E-waste with a generation rate of 24.9 Mt which amounts to a per capita generation rate of 5.6 kg/capita-year. In comparison, Europe had the highest per capita generation rate of 16.2 kg/capita-year. It is estimated that only 17.4% of global e-waste is formally collected and recycled.

Table 2 Typical weights and life spans of E-waste (Premalatha et al. 2014)

Item	Mass (kg)	Typical life span (years)
Computer	25	3
Television	30	5
Mobile telephone	0.1	2
Video recorder and DVD player	5	5
Radio	2	10
Photocopier	60	8
Telephone	1	5
Refrigerator	35	10
Air conditioning unit	55	12
Electric heaters	5	20
Washing machine	65	5
Electric cooker	60	10
Iron	1	10
Food mixer	1	5
Microwave	15	7
Hair dryer	1	10
Vacuum cleaner	10	10
Toaster	1	5

Non-ferrous metals include copper, aluminium, and precious metals like gold, silver, platinum, palladium, etc. Heavy metals like lead, mercury, arsenic, cadmium, etc. are also present in large quantities

A list of the top ten E-waste generating countries in 2014 and 2020 in order of the annual E-waste generated is provided in Table 4a, b, respectively (Baldé et al. 2015; Forti et al. 2020a, b). India was the fifth largest generator of E-waste at the global level in 2014 and in 2020, it was the third largest e-waste generator in the world. However, per capita, E-waste generation rates in India are amongst the lowest (1.3 kg/capita-y) in comparison to the highest per capita E-waste generation rates of 28.3 kg/capita-y in Norway.

In the global scenario, the demand for certain metals like rare earth metals, platinum group elements, certain base metals, and precious metals, etc. has increased day-by-day but the natural resources are either not available or are diminishing in quantity. Thus, one option is to recover metals from e-waste in a process known as 'Urban mining'. Urban mining is the process of recovering different compounds and elements from various wastes which would otherwise be left open in the environment to cause different levels of pollution (Fig. 1). By changing our way of thinking, we can reuse and recycle valuable components from the wastes, paving the way towards a zero-landfill future and can also save our environment.

Table 3 (continued)

E-waste	Constituent	Effect on human health	Effect on environment	References
Light emitting diodes (LEDs)	Arsenic (As) in the form of gallium arsenide	Acutely poisonous and injurious to health		Begum (2013)
Li-batteries	Lithium (Li)		May produce explosive gases (hydrogen) if wetted	Begum (2013)
Rechargeable Ni–Cd batteries or NiMH batteries, electron gun in CRT	Nickel (Ni)	May cause allergic reactions		Begum (2013)
Fluorescent layer in CRT-screen	Rare earth elements (Yttrium, Europium, Neodymium, etc.)	Irritates skin and eyes		Begum (2013)
CRT-screen	Zinc sulphide	Toxic when inhaled		Begum (2013)
Condensers, liquid crystal display	Toxic organic substances			Begum (2013)
Toner cartridges for laser printers/ copiers	Toner dust	Health risk when dust is inhaled		Begum (2013)

Table 4a Top ten E-waste generating countries in the world in 2014 (Baldé et al. 2015)

Rank	Country	Continent	E-waste generation rate, kg/person-y	E-waste generation rate, Kilo tons/y	National regulations in effect, 2013
1	USA	Americas	22.1	7,072	No
2	China	Asia	4.4	6,033	Yes
3	Japan	Asia	17.3	2,200	Yes
4	Germany	Europe	21.6	1,769	Yes
5	India	Asia	1.3	1,641	No
6	UK and N Ireland	Europe	23.5	1,511	Yes
7	France	Europe	22.1	1,419	Yes
8	Brazil	Americas	7	1,412	No
9	Russia	Europe	8.7	1,231	No
10	Italy	Europe	17.6	1,077	Yes

Table 4b Top ten E-waste generating countries in the world in 2020 (Forti et al. 2020a, b)

Rank	Country	E-waste generated 2019	E-waste generated 2019	E-waste documented as collected and recycled	Year documented	National e-waste legislation/policy or regulation in place
		(kilotons)	kg/cap-y	kilotons		
1	China	10129	7.2	1546	2018	Yes
2	USA	6918	21	1020	2017	Yes
3	India	3230	2.4	30	2016	Yes
4	Japan	2569	20.4	570	2017	Yes
5	Brazil	2143	10.2	0.14	2012	No
6	Russian Federation	1631	11.3	90	2014	No
7	Indonesia	1618	6.1	NA	2019	No
8	Germany	1607	19.4	837	2017	Yes
9	UK	1598	23.9	871	2017	Yes
10	France	1362	21	742	2017	Yes

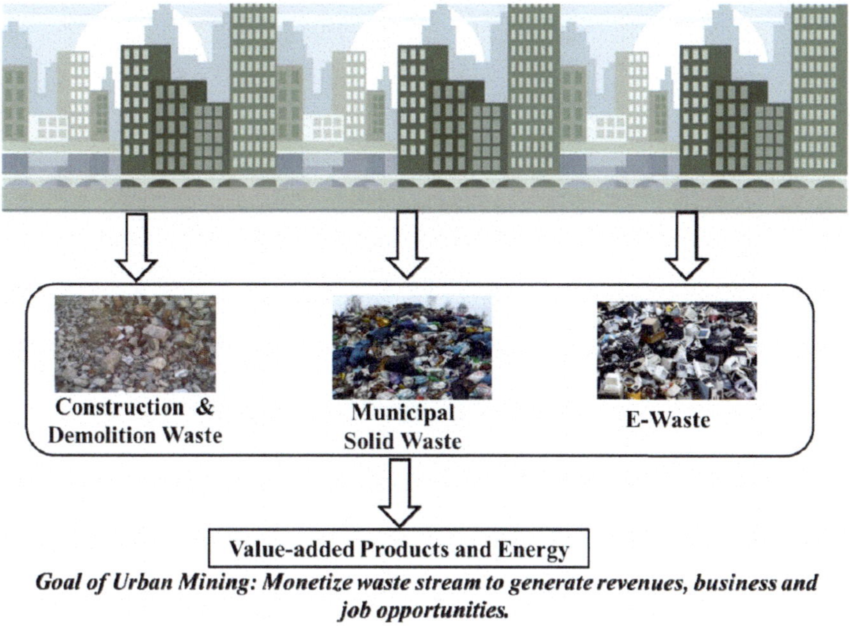

Fig. 1 Management of e-waste through urban mining

3.2 Indian Scenario of E-Waste Generation and Management

The most recent estimate (2019) of E-waste generation in India is 3.23 million metric tons/y (Forti et al. 2020a, b). CPCB's estimates of e-waste generation, dismantling and recycling rates are shown in Table 5 and India now has 472 registered E-waste recyclers in 22 states as of 29 April 2022 with a total e-waste handling capacity of 1,426,685.22 metric tons/annum (CPCB 2021, 2022).

The top ten states in India generating E-waste are shown in Table 6. These 10 states generate 70% of the total E-waste and 65 cities are responsible for generating 60% of the total E-waste in the country (Vats and Singh 2014). Issues regarding the collection of information from industry and compliance with regulations are major deterrents in E-waste management in India.

Table 5 E-waste generation and recycling rates in India in metric tons/year (CPCB 2021)

Year	E-waste generation	E-waste dismantling and recycling	% recycled
2017–18	7,08,445	69,414	9.80
2018–19	7,71,215	1,64,663	21.35
2019–20	10,14,961	2,22,436	21.92

Table 6 E-waste/WEEE generation in ten states of India (Vats and Singh 2014)

Rank	States	WEEE (tons)	Percentage (%)
1	Maharashtra	20,270.59	18.49
2	Tamil Nadu	13,486.24	12.30
3	Andhra Pradesh	12,780.33	11.d6
4	Uttar Pradesh	10,381.1 I	9.47
5	West Bengal	10,059.36	9.1
6	Delhi	9729.15	8.87
7	Karnataka	9115.74	5.32
8	Gujarat	8994.33	8.20
9	Madhya Pradesh	7500.62	7.11
10	Punjab	6955.46	6.35
Total		109,578.93	100

4 Management Practices for E-Waste

For the proper management of E-waste, effective and practical solutions are required. The first principle is to reduce the generation rate of E-waste at the source. Second, the number of collection centers should be increased in every state. After collection, wastes should be transported safely to recycling centers for further processing. Various rules, regulations and methods are adopted for the management of e-waste (Fig. 2).

4.1 E-Waste (Management and Handling) Rules

The E-waste (Management and Handling) Rules, 2011 were amended in the year 2016 and are now called "The E-waste (Management and Handling) Rules, 2016". These rules are effective since 1 Oct 2016. They were further amended on 22 March 2018.

The rules apply to every manufacturer, producer, consumer, bulk consumer, collection centres, dealers, refurbishers, dismantlers and recyclers involved in the manufacture, sale, transfer, purchase, collection, storage and processing of E-waste or

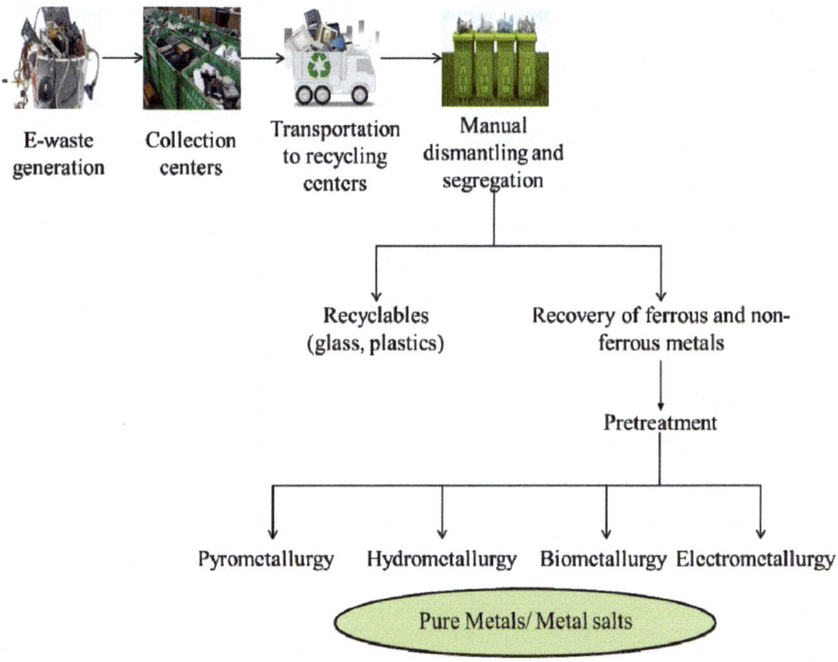

Fig. 2 Steps involved in E-waste management

electrical and electronic equipment (EEE). It also explains the role of the producer clearly. Producer Responsibility Organizations (PROs) are also included in the new rules of 2016 in addition to the Extended Producer Responsibility (EPR). Penalty and punishment for non-compliance are defined in Sections 15 and 16 of the Environment (Protection) Act, 1986 which says that whoever fails to comply with the rules will be punishable with imprisonment for a certain period which may be extended to 5 years or with a penalty which may extend to Rs 1 lakh. The E-waste rules were again amended in the year 2018 as E-waste (Management) Amendment Rules, 2018. The salient features of these amended rules are as follows.

Some of the salient features of the Amendment Rules are:

- The collection targets for E-waste under EPR have been revised with effect from 01-10-2017. The phase wise collection targets for E-waste in weight shall be 10% of the quantity of waste generation as indicated in the EPR Plan during 2017–18, with a 10% increase every year until 2023. From 2023 onwards, the target has been made 70% of the quantity of waste generation as indicated in the EPR Plan.
- The quantity of e-waste collected by producers from 01-10-2016 to 30-09-2017 shall be accounted for in the revised EPR targets until March 2018.
- Separate E-waste collection targets have been drafted for new producers, i.e. those producers whose number of years of sales operation is less than the average lives of their products. The average lives of the products will be as per the guidelines issued by CPCB from time to time.
- Producer Responsibility Organizations (PROs) shall apply to the Central Pollution Control Board (CPCB) for registration to undertake activities prescribed in the Rules.
- Under the Reduction of Hazardous Substances (RoHS) provisions, cost for sampling and testing shall be borne by the government for conducting the RoHS test. If the product does not comply with RoHS provisions, then the cost of the test will be borne by the producers.

4.2 Extended Producer Responsibility (EPR)

EPR is an environmental protection strategy that makes the producer responsible for the entire life cycle of the product and principally for take-back, recycling, and final disposal. EPR uses financial incentives to encourage manufacturers to design environmental-friendly products and hold them responsible for the costs of managing their products to the end-of-life (EOL). A summary of EPR legislation in different countries is provided in Table 7.

Table 7 The role of EPR and E-waste legislation in different countries

Countries	EPR concept
European Union	The directive mainly focuses on reuse, recycle and recovery of E-wastes and dismantling of electronic parts and recycling of materials, proper collection systems to reduce disposal and incorporate best management practices
Switzerland	This country was the first to develop and implement methods for collection, transportation, recycling/treatment, and disposal of E-waste. Three producer responsibility organizations (PROs): The Swiss Association for Information Communication and Organizational Technology (SWICO); the Stiftung Entsorgung Schweiz (SENS); and Swiss Lighting Recycling Foundation (SLRS) oversee these systems based on the concept of EPR
The United States of America	Established funding for the collection and recycling of E-waste. Consumers pay a fee called Advance Recycling Fee (ARF) at the time of purchase that goes to the state and is used to reimburse recyclers and collectors
Japan	Manufacturers and importers are responsible for taking back end-of-life electronics for recycling and waste management. Consumers pay a fee that is directly used to meet the expenses of recycling and transportation
South Korea	Local manufacturers, distributors and importers of e-goods are required to achieve official recycling targets. Government keeps an account for depositing funds for recycling, which are refundable depending on the amount of waste recycled
Australia	Importers, manufacturers, and distributors have to subscribe to mandatory, co-regulatory, or voluntary schemes for managing the disposal of computers and televisions
Singapore	Export, import or transit waste requires a permit from the Pollution Control Department (PCD) of Singapore. If documents are available to support that the products are in good condition and can be reused, only then permission is granted
China	Pollution prevention and controls on the use, dismantling and disposal of E-waste, under "Technical Policies for Controlling Pollution of WEEE, 2006", Certificate is required for E-waste recycling systems, under "Administrative Measures for the Prevention and Control of Environmental Pollution by WEEE, 2005", and All producers and importers are responsible for their products, collection, and treatment funds, under "Regulation on the Administration of the Recovery and Disposal of WEEE, 2011"
India	According to MoEF, 2011, producers are responsible for the collection of E-waste generated after the end-of-life of the E-products. The legislation has been modified in 2015 with new responsibilities for the producers as well as consumers
African countries	The use of e-goods is very less in African countries. No specific e-waste legislation has been implemented in those countries

Source Premalatha et al. (2014)

or disposal. LCD screens in television sets contain several thin mercury lamps and liquid crystals which makes them a challenge for recycling.

5.2 *Plastic*

All plastics are transferred to recyclers who use them as raw materials to manufacture new products without altering their properties. However, there is a big challenge in separating the high-value plastic streams from the mixed plastics present in E-waste. The CreaSolv process has been developed for the recovery of polymer fraction from the plastics present in mobile phones (Maurer et al. 2004). Plastic recovery techniques are difficult due to the variety of polymers used in e-goods and the labour-intensive nature of the separation of plastic from E-waste.

5.3 *Ferrous Metals and Non-ferrous Metals*

Magnetic separation of ferrous metals and non-ferrous metals can be done to separate the two fractions. Recovery of various non-ferrous metals can be done by using any of the following processes (Rajesh et al. 2022; Hsu et al. 2019; Awasthi et al. 2019, 2022; Ambaye et al. 2020; Yanamandra et al. 2022):

1. Pyrometallurgy
2. Hydrometallurgy
3. Biometallurgy or Bioleaching
4. Electrometallurgy.

5.3.1 Pyrometallurgy

Pyrometallurgy is the thermal treatment of minerals and metallurgical ores resulting in the physical and chemical transformation of valuable metals. In this process, E-waste is heated above 1500 °C in a closed reactor, i.e., in the absence of oxygen, specially designed to ensure zero emissions. This process includes several procedures like calcining, roasting, smelting and refining. Elements that can be recovered by the pyrometallurgical process include oxides of less reactive elements like iron, zinc, copper, manganese, chromium, tin, etc. Some pyrometallurgical studies for metal recovery are summarized in Table 8.

5.3.2 Hydrometallurgy

Appropriate chemical reagents are used in hydrometallurgy to dissolve metals from E-waste. The various chemical reagents used are aqueous and organic solvents,

Table 8 Pyrometallurgical methods for recovery of metals from E-waste

Techniques	Metals recovered	Main results	References
Aleksandrovich's patent for recovery of gold and platinum group metals from electronic scraps	PGM and gold	PGM and gold were recovered	Aleksandrovich et al. (1998)
Noranda process, Quebec, Canada	Cu, Au, Ag, Pt, Pd, Se, Te, Ni	Recovery of base metal like copper and precious metals	Khaliq et al. (2014), Kaya (2016)
Boliden Rönnskår Smelter, Sweden	Cu, Ag, An, Pd, Ni, Se, Zn, Pb	Recovery of copper and precious metals	Khaliq et al. (2014), Kaya (2016)
Umicore's precious metal refining process, Hoboken, Belgium	Base metals, precious metals, platinum group metals and selenium, tellurium, indium	Almost complete recovery of base metals, precious metals and special metals such as Sb, Bi, Sn, Se, Te, In	Khaliq et al. (2014), Kaya (2016)
Thermal treatment	Phosphate and rare earth metals (REMs) from Korean monazite ore	>90% REMs	Kumari et al. (2015)
Aurubis, Lunen, Germany	Cu, Au, Ag, Pb, Zn, Sn		Khaliq et al. (2014)
DOWA smelter, Kosaka, Japan	Cu, Au, Ag, Ni, Bi		Kaya (2016)

mineral acids, some bases, mixed acids, etc. The various types of solvents used are halides, cyanides, thiourea, and thiosulfate for the recovery of precious metals from E-waste (Tembhare et al. 2022). The different processes involved in hydrometallurgy are:

(a) Leaching: The most efficient leaching agents are acids as they have the capacity to solubilize both, base and precious metals.
(b) Solution concentration and purification: The main purpose is to remove undesired impurities so that metal concentration in the solution increases. Purification of the solution is done by ion exchange, adsorption, and solvent extraction processes.
(c) Recovery of metal: After solution purification, metal is to be recovered chemically or electrochemically.

Applications of hydrometallurgical methods for metals recovery from E-waste are summarized in Table 9.

Table 9 Hydrometallurgical methods for recovery of metals from E-waste

Materials	Metals recovered	Efficiency	References
Computer chips	Au	Gold flakes	Sheng and Etsell (2007)
Ceramic capacitors	Ni	Ni in solution	Kim et al. (2007)
E-waste	Au, Pd, Pt, Ag	Au-98%, Pd-969%; Pt-92%; Ag- 84%	Kogan (2006)
E-waste	Au, Ag, Pd	92% for Au, Ag, Pd	Zhou et al. (2005)
E-waste	Cu, Pb, Sn	Cu, Pb, Sn	Mecucci and Scott (2002)
E-waste	Au, Ni	Au, Ni	Zakrewski et al. (1992)
Solder of waste printed circuit boards	Lead	Pb-99.99%	Kumari et al. (2010)
Printed circuit boards	Cu, Fe, Ni, Pb	Cu, Fe, Ni-99.99% Pb-36.7%	Jha et al. (2010)
Ga-As waste scraps	Ga	Ga-50.8%	Hu et al. (2015)
Printed circuit boards	Ag	Ag (97%)	Neto et al. (2016)
NiMH batteries	La, Nd, Sm, Pr, Ce	La-2.7% Nd-4.5% Sm-6.2% Pr-2.29% Ce-2.58%	Marra et al. (2018)
Leaching with HNO_3	Cu		Dutta et al. (2018a)
Leaching using H_2SO_4 and H_2O_2	Co, Li, Cu		Dutta et al. (2018b)
Leaching with HCl	Cu, Au		Li et al. (2019)
Leaching using air + H_2SO_4	Cu		Guo et al. (2020)
Leaching with NH_4NO_3 + $Na_2S_2O_3$	Au, Ag		Mishra et al. (2021)

5.3.3 Biometallurgy

Biometallurgy involves the use of microbes in an aqueous environment to produce metals and has been used for decades for the recovery of metals like copper, gold, and uranium. Almost a quarter of the global production of copper is done by bioleaching (Madigan et al. 2015). Bioleaching is defined as the mobilization of metal cations from mineral ores by biological oxidation. Biometallurgical methods used for metal recovery are summarized in Table 10.

Table 10 Biometallurgical methods for metal recovery from E-waste

Materials	Metals recovered %	Microorganism used	References
Ni–Cd battery	Cd-54	*A. ferrooxidans* and *A. thiooxidans*	Velgosova et al. (2010)
E-waste	Ni (89.4%), Co (50.4%), Li (99.2%)	*A. thiooxidans* and *A. ferrooxidans*	Heydarian et al. (2018)
E-waste	Cu (90%) and Ni (89%)	*Penicillium simplicissimum*	Arshadi et al. (2019)
Waste PCBs	Au-98 Ag-14	*Acidiplasma* sp.	Rizki et al. (2019)
E-waste	Au (100% and 34% respectively)	*Roseovarius tolerans and Roseovarius mucosus*	Kudpeng et al. (2020)
Waste PCBs	(87.5% Cu and 73.6% Au) and (72.7% Cu and 66.6% Au)	*C. violaceum* and *B. megaterium* respectively	Kumar et al. (2021)
Spent mobile phone PCBs	Cu (98%) and Ni (82%)	*A. thiooxidans*	Kadivar et al. (2021)

5.3.4 Electrometallurgy

Electrometallurgy involves the use of electro-plating or electro-deposition methods. In these methods, electrolytic cells with different types of electrolytes and electrodes are used for the deposition of expensive metals on the surface of the appropriate electrode. The most common processes used in electrometallurgy are.

(a) Electro-winning: A leach solution containing the metal of interest is added to an electrolytic cell. When current is passed through the cell, the metal is deposited on the cathode while the anode remains inert.
(b) Electro-refining: An acidic electrolyte is added to an electrolytic cell with an anode made of impure metal. When current is passed through the cell, the anode dissolves into the electrolyte from where pure metal is deposited on the cathode resulting in greater metal purity.
(c) Electro-forming: Thin metal parts are moulded on pre-formed cathodes and manufactured through electroplating.

6 Conclusions

E-waste management has become a serious concern for the world due to the exponential increase in the use of electrical and electronic products and the resultant waste generated. Best methods from different countries can be incorporated to develop and design safe and hazard-free E-waste management systems. Proper E-waste management can help in efficient resource recovery and in E-waste management from cradle-to-grave. In India, E-waste (Management and Handling) Rules, 2011 have been modified to E-waste (Management) Rules, 2016, in which producers, distributors, and importers are held responsible for the proper recycling and disposal of E-waste. Efficient collection of E-waste and its treatment can lead to a better environment and huge piles of E-waste can be turned into lucrative products and business opportunities.

References

Aleksandrovich, S., Nicolaevich, E. and E. Ivanovich (1998). Method of processing of products based on chalcogenides of base metals containing metals of platinum group and gold. Russian Patent, RU2112064, C22B 11/02.

Ambaye, T.G., Vaccari, M., Castro, F.D., Prasad, S. and S. Rtimi (2020). Emerging technologies for the recovery of rare earth elements (REEs) from the end-of-life electronic wastes: A review on progress, challenges, and perspectives. *Environmental Science and Pollution Research,* **27**(**29**): 36052–36074. https://doi.org/10.1007/s11356-020-09630-2.

Andreola, F., Barbieri, L., Corradi, A. and L. Lancellotti (2005). Cathode ray tube glass recycling: An example of clean technology. *Waste Management & Research,* **23**: 314–321.

Arshadi, M., Nili, S. and S. Yaghmaei (2019). Ni and Cu recovery by bioleaching from the printed circuit boards of mobile phones in non-conventional medium. *Journal of Environmental Management*, **250**: 109502.

Awasthi, A.K., Li, J., Koh, L. and O.A. Ogunseitan (2019). Circular Economy and Electronic Waste. *Nature Electronics*, **2**(**3**): 86–89. https://doi.org/10.1038/s41928-019-0225-2.

Awasthi, A.K., Awasthi, M.K., Saket Mishra, S., Sarsaiya, S. and A.K. Pandey (2022). Evaluation of e-waste materials linked potential consequences to environment in India. *Environmental Technology and Innovation*, **28**: 102477. https://doi.org/10.1016/j.eti.2022.102477.

Baldé, C.P., Wang, F., Kuehr, R. and J. Huisman (2015). The Global E-Waste Monitor 2014. UNU-IAS: Bonn, Germany.

Begum, K.J. (2013). Electronic waste (e-waste) management in India: A review. *IOSR Journal of Humanities and Social Science (IOSR-JHSS),* **10**(**4**): 46–57.

CPCB (2021). CPCB Annual Report 2020–2021. *Annual Report 2020–2021.* Delhi, India. https://www.pvh.com/-/media/Files/pvh/investor-relations/PVH-Annual-Report-2020.pdf.

CPCB (2022). List of Dismantlers / Recyclers. Delhi, India: CPCB. https://cpcb.nic.in/.

Dutta, D., Panda, R., Kumari, A., Goel, S. and M.K. Jha (2018a). Sustainable recycling process for metals recovery from used printed circuit boards (PCBs). *Sustainable Materials and Technologies,* **17**: e00066. https://doi.org/10.1016/j.susmat.2018.e00066.

Dutta, D., Kumari, A., Panda, R., Jha, S., Gupta, D., Goel, S. and M.K. Jha (2018b). Close loop separation process for the recovery of Co, Cu, Mn, Fe and Li from spent lithium-ion batteries. *Separation and Purification Technology*, **200**: 327–334. https://doi.org/10.1016/j.seppur.2018.02.022

Dutta, D. and S. Goel (2021). Understanding the gap between formal and informal e-waste recycling facilities in India. *Waste Management*, **125**: 163–171.

Forti, V., Baldé, C.P., Kuehr, R. and G. Bel (2020). The Global E-Waste Monitor 2020. Bonn/Geneva/Rotterdam: Flows and the circular economy potential. United Nations University (UNU)/United Nations Institute for Training and Research (UNITAR).

Forti, V., Baldé, C.P., Kuehr, R. and G. Bel (2020). *The Global E-Waste Monitor 2020.* UNU, UNITAR, ITU and ISWA. http://ewastemonitor.info/.

Fu, J., Zhong, J., Chen, D. and Q. Liu (2020). Urban environmental governance, government intervention, and optimal strategies: A perspective on electronic waste management in China. *Resources, Conservation and Recycling*, **154**: 104547.

Guo, X., Qin, H., Tian, Q. and D. Li (2020). Recovery of metals from waste printed circuit boards by selective leaching combined with cyclone electrowinning process. *Journal of Hazardous Materials*, **384**: 121355. https://doi.org/10.1016/j.jhazmat.2019.121355

Heydarian, A., Mousavi, S.M., Vakilchap, F. and M. Baniasadi (2018). Application of a mixed culture of adapted acidophilic bacteria in two-step bioleaching of spent lithium-ion laptop batteries. *Journal of Power Sources*, **378**: 19–30.

Hsu, E., Barmak, K., West, A.C. and A.H.A. Park (2019). Advancements in the treatment and processing of electronic waste with sustainability: A Review of metal extraction and recovery technologies. *Green Chemistry*, **21(5)**: 919–936. https://doi.org/10.1039/c8gc03688h.

Hu, S.H., Xie, M.Y., Hsieh, Y.M., Liou, Y.S. and W.S. Chen (2015). Resource recycling of gallium arsenide scrap using leaching-selective precipitation. *Environmental Progress & Sustainable Energy*, **34(2)**: 471–475. https://doi.org/10.1002/ep.12019

Jha, M.K., Kumar, S.V., Pandey, B.D., Kumar, R. and J. Lee (2010). Leaching studies for the recovery of metals from the waste Printed Circuit Boards (PCBs). *The Minerals, Metals & Materials Society,* Proceedings of TMS-2010, pp. 945–952.

Kadivar, S., Pourhossein, F. and S.M. Mousavi (2021). Recovery of valuable metals from spent mobile phone printed circuit boards using biochar in indirect bioleaching. *Journal of Environmental Management*, **280**: 111642.

Kaya, M. (2016). Recovery of metals and non-metals from electronic waste by physical and chemical recycling processes. *Waste Management*, **57**: 64–90. https://doi.org/10.1016/j.wasman.2016.08.004.

Khaliq, A., Rhamdhani, M.A., Brooks, G. and S. Masood (2014). Metal extraction processes for electronic waste and existing industrial routes: A review and Australian perspective. *Resources*, **3(1)**: 152–179. https://doi.org/10.3390/resources3010152.

Kim, E., Lee, J. and B.S. Kim (2007). Leaching behavior of nickel from waste multi-layer ceramic capacitors. *Hydrometallurgy,* **86(1–2)**: 89–95.

Kogan, V. (2006). Process for the recovery of precious metals from electronic scrap by hydrometallurgical technique. International Patent, WO/2006/013568 (C22B 11/00), W.I.P. Organization.

Kudpeng, K., Bohu, T., Morris, C., Thiravetyan, P. and A.H. Kaksonen (2020). Bioleaching of gold from sulfidic gold ore concentrate and electronic waste by *Roseovarius tolerans* and *Roseovarius mucosus*. *Microorganisms,* **8(11)**: 1783.

Kumar, A., Saini, H.S., Şengör, S., Sani, R.K. and S. Kumar (2021). Bioleaching of metals from waste printed circuit boards using bacterial isolates native to abandoned gold mine. *BioMetals*, **34(5)**: 1043–1058.

Kumari, A., Jha, M.K., Kumar, V., Jeong, J. and J. Lee (2010). Recovery of lead from the solder of waste Printed Circuit Boards. Proceedings of the XI International Seminar on Mineral Processing Technology (MPT-2010).

Kumari, A., Panda, R., Jha, M.K., Lee, J.Y., Kumar, J.R. and V. Kumar (2015). Thermal treatment for the separation of phosphate and recovery of rare earth metals (REMs) from Korean monazite. *Journal of Industrial and Engineering Chemistry,* **21**: 696–703.

Li, F., Chen, M., Shu, J., Shirvani, M., Li, Y., Sun, Z., Sun, S., Xu, Z., Fu, K. and S. Chen (2019). Copper and gold recovery from CPU sockets by one-step slurry electrolysis. *Journal of Cleaner Production*, **213**: 673–679. https://doi.org/10.1016/j.jclepro.2018.12.161.

Madigan, M.T., Martinho, J.M., Bender, K., Buckley, D. and D. Stahl (2015). *Brock Biology of Microorganisms.* Pearson Publishers.

Maphosa, M. and V. Maphosa (2022). A bibliometric analysis of the effects of electronic waste on the environment. *Global Journal of Environmental Science and Management*, **8(4)**: 589–606. https://doi.org/10.22034/GJESM.2022.04.10.

Marra, A., Cesaro, A. and V. Belgiorno (2018). Separation efficiency of valuable and critical metals in WEEE mechanical treatments. *Journal of Cleaner Production*, **186**: 490–498. https://doi.org/10.1016/j.jclepro.2018.03.112.

Maurer, A., Schlummer, M. and G. Altnau (2004). Material recovery from mobile phones by CreaSolv process. *In:* Electronic Goes Green. Fraunhofer IZM, Berlin.

Mecucci, A. and K. Scott (2002). Leaching and electrochemical recovery of copper, lead and tin from scrap printed circuit boards. *Journal of Chemical Technology and Biotechnology,* **77(4)**: 449–457.

Mishra, G., Jha, R., Rao, M.D., Meshram, A. and K.K. Singh (2021). Recovery of silver from waste printed circuit boards (WPCBs) through hydrometallurgical route: A review. *Environmental Challenges*, **4**: 100073. https://doi.org/10.1016/j.envc.2021.100073.

Neto, I.F.F., Sousa, C.A., Brito, M.S.C.A., Futuro, A.M. and H.M.V.M. Soares (2016). A simple and nearly closed cycle process for recycling copper with high purity from end life printed circuit boards. *Separation and Purification Technology*, **164**: 19–27. https://doi.org/10.1016/j.seppur.2016.03.007.

Premalatha, M., Abbasi, T., Abbasi, T. and S.A. Abbasi (2014). The generation, impact, and management of e-waste : State of the art the generation, impact, and management. *Critical Reviews in Environmental Science and Technology*, **44(14)**: 1577–1678. https://doi.org/10.1080/10643389.2013.782171.

Rajesh, R., Dharmaraj, K. and N. Prabaharan (2022). Electronic waste: A critical assessment on the unimaginable growing pollutant, legislations and environmental impacts. *Environmental Challenges*, **7**: 100507. https://doi.org/10.1016/j.envc.2022.100507.

Ramachandra, T.V. and K. Saira Varghese (2004). Environmentally sound options for e-wastes management. *Envis Journal of Human Settlements,* March.

Rizki, I.N., Tanaka, Y. and N. Okibe (2019). Thiourea bioleaching for gold recycling from e-waste. *Waste Management*, **84**: 158–165.

Sankhla, M.S., Kumari, M., Nandan, M., Mohril, S., Singh, G.P., Chaturvedi, B. and R. Kumar (2016). Effect of electronic waste on environmental & human health-a review. *IOSR Journal of Environmental Science, Toxicology and Food Technology*, **10**: 98–104.

Sheng, P.P. and T.H. Etsell (2007). Recovery of gold from computer circuit board scrap using aqua regia. *Waste Management Research,* **25(4)**: 380–383.

Singh, V., Cortes-Ramirez, J., Toms, L.M., Sooriyagoda, T. and S. Karatela (2022). Effects of polybrominated diphenyl ethers on hormonal and reproductive health in e-waste-exposed population: A systematic review. *International Journal of Environmental Research and Public Health*, **19(13)**: 7820. https://doi.org/10.3390/ijerph19137820.

Tembhare, S.P., Bharat, A.B., Barai, D.P. and S.J. Dhoble (2022). E-Waste Recycling Practices: A Review on Environmental Concerns, Remediation and Technological Developments with a Focus on Printed Circuit Boards. *In:* Environment, Development and Sustainability. Vol. 24. Springer Netherlands. https://doi.org/10.1007/s10668-021-01819-w.

Terazono, A., Murakami, S., Abe, N., Inanc, B., Moriguchi, Y., Sakai, S.I., Kojima, M., Yoshida, A., Li, J., Yang, J. and M.H. Wong (2006). Current status and research on E-waste issues in Asia. *Journal of Material Cycles and Waste Management*, **8(1)**: 1–12.

UNEP (2015). Report of the Conference of the Parties to the Basel Convention on the Control of Transboundary Movements of Hazardous Wastes and Their Disposal on the work of its twelfth meeting, Geneva.

Vats, M.C. and S.K. Singh (2014). Status of e-waste in India—A review. *International Journal of Innovative Research in Science, Engineering and Technology*, **3(10)**: 16917–16931. https://doi.org/10.15680/IJIRSET.2014.0310071.

Velgosova, O., Kadukova, J., Mrazikova, A., Blaskova, A., Petoczova, M., Harvathova, H. and M. Stofko (2010). Influence of selected parameters on nickel bioleaching from spent nickel-cadmium batteries. *lVtineralia Slovaca,* **42**: 365–368.

Yanamandra, K., Pinisetty, D., Daoud, A. and N. Gupta. (2022). Recycling of Li-ion and lead acid batteries: A review. *Journal of the Indian Institute of Science*, **102**(**1**): 281–95. https://doi.org/10.1007/s41745-021-00269-7.

Zakrewski, J., Chamer, R. and A. Koscielniak (1992). Method for recovering gold and nickel from electronic scrap and waste. Polish Patent, PL15 b889B (C22B7/00).

Zhou, P., Zheng, Z. and J. Tie (2005). Technological process for extracting gold, silver and palladium from electronic industry waste. *Chinese Patent,* CN1603432A (C22B 11/00).

Chapter 3
Survey of Composting Companies and Compost Samples in India

B. R. Hiremath and Sudha Goel

1 Introduction

The unwanted, useless or used solid materials generated from combined residential, industrial and commercial activities are known as solid waste. As per Solid Waste Management Rules, 2016, garbage is defined as Municipal Solid Waste which includes commercial and residential wastes generated in municipal or notified areas in either solid or semi-solid form excluding industrial hazardous wastes but including treated bio-medical wastes. Municipal solid waste consists of household waste, construction and demolition debris, sanitation residue, and waste from the streets.

The Census of 2011 estimates India's population to be 1.21 billion which is 17.66% of the world population. Increasing population levels, rapid economic growth and a rise in community living standards have accelerated the generation rate of municipal solid waste (MSW) in Indian cities. India's urban population was 285 million in 2001 and increased to 377 million in 2011. It is estimated that 152,077 tons/day of MSW were generated in the urban centers in India (CPCB 2013). Of this waste, 98.5% was collected, 36.7% was treated and 33% was disposed off in landfills. The solid waste generated in any area is generally directly dependent on the economy and the percentage of the urban population.

Organic waste composting is one of the best methods for treating MSW and is a viable alternative to land-filling and incineration (Ribeiro et al. 2000; Rosen et al. 1993; Manios 2004; Adani et al. 2004). Aerobic composting is defined as the decomposition of organic materials in the presence of oxygen. Since composting

B. R. Hiremath (✉)
Civil Engineering Department, Basaveshwar Engineering College, Bagalkot,
Karnataka 587 102, India
e-mail: brhiremath1@gmail.com

S. Goel
Civil Engineering Department and School of Environmental Science and Engineering, Indian
Institute of Technology Kharagpur, Kharagpur 721 302, India

S. Goel (ed.), *Advances in Solid and Hazardous Waste Management*,
https://doi.org/10.1007/978-3-031-49144-3_3

Table 1 Compost samples and their description

Sl. No	Composts	Description
1	BHR	Bhadreswar Municipal Solid Waste Treatment plant Compost (Vermi-compost)
2	DHP	Dhapa Landfill Compost (Windrow)
3	BGK 1	Bagalkot Compost 1
4	BMC	Bone Meal Compost
5	NPC	Neem Patta Compost
6	AGR	Agriculture and Food Engineering Department Compost
7	BGK 2	Bagalkot Compost 2 (Vermi-compost)
8	AHD	Ahmedabad Compost (Compost)
9	HC	Kharagpur Home Compost
10	OC 1	Orissa Compost 1
11	OC 2	Orissa Compost 2

leads to a stabilized final product, which will improve soil quality and fertility, it is one of the best options for organic waste disposal among the various strategies for waste management. The specific carbon-to-nitrogen ratio and the moisture content may increase the effectiveness of the process of composting (Chang and Chen 2010; Kumar et al. 2010). Factors like feed composition, moisture content, oxygen content, temperature, and pH affect microbial activities and therefore, the composting process (Bhatia et al. 2013). Bulking agents like rice husk, sawdust and rice bran increase the quality of compost derived from food waste (Chang and Chen 2010). Urea addition in crop waste was in practice in the past to reduce the C/N ratio (Gaind and Gaur 2000). However, in recent practice, nitrogen-rich organic wastes such as poultry litter and cow dung wastes (C/N ratio <12) are used in organic farming as supplements to increase the nitrogen content of compost (Kalemelawa et al. 2012).

A survey was conducted of composting companies in India using the Internet and the methods adopted commercially. Furthermore, a quantitative and qualitative analysis of eleven different compost samples from different parts of India was done based on laboratory elemental analysis. Eleven compost samples were also collected and tested for their elemental composition and their description is given in Table 1.

2 Materials and Methods

2.1 Data Collection

On 16 August 2013, a list of composting companies in India with their Internet addresses was obtained from www.compost-india.com. These composting companies were contacted by phone for information regarding their operations and compost production. However, no company was willing to provide any information over the phone or during visits. Therefore, data collection for the survey was limited to the information provided by the companies on their websites. Of the 181 companies listed on www.compost-india.com, only 101 companies provided details on their websites. Information collected for analysis of composting companies and their products included the following factors: number of composting companies in each state, raw materials used, cost per kg of compost, compost quality if it was tested, and method of composting. Information from the individual websites was copied manually into MS Excel spreadsheets.

2.2 *Sample Collection*

Further, eleven compost samples were collected from different locations all over the country and tested for various parameters. A description of the samples and types of compost is provided in Table 1. Single compost samples were collected in plastic bags and sample sizes varied from 1 to 5 kg.

2.3 *Analytical Methods*

Moisture content, bulk density, total solids (TS), volatile solids (VS), and fixed solids (FS) were measured according to standard methods. Scanning electron microscope (ZEISS EVO 60, Carl ZEISS SMT, Germany) images were used to obtain information about the sample's surface topography. Further, Energy Dispersive X-ray Spectroscopy (EDX) was used to determine elements on the surface of the sample for qualitative information about compost composition. C, H, N, S and O elements were determined in the compost samples using a CHNSO analyser (Eurovector, Italy).

3 Results and Discussion

3.1 *Composting Companies in India*

The Internet survey of composting companies resulted in data for 101 companies in India and their state-wise distribution is shown in Fig. 1. These companies were distributed in 14 states and the largest number of companies was in Karnataka (26). Compost companies were selling their composts for different purposes like soil conditioner, mulch, growing media, and improving topsoil composition. Different types of composts like commercial compost (compost that is sold in the market), vermicompost, farm/manure compost, mixed municipal compost, and forest/wood-chip compost were being manufactured by these companies. Most of the compost companies were using green waste or kitchen waste as their raw material (53%) while some were using plant waste and animal manure as shown in Fig. 2.

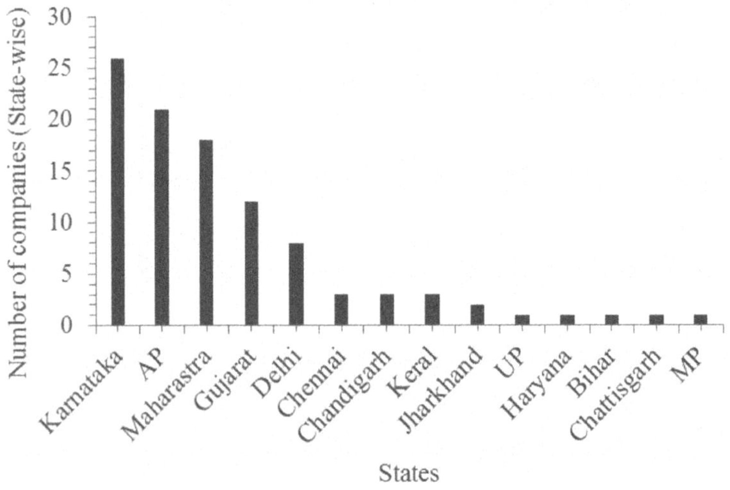

Fig. 1 State-wise distribution of composting companies in India

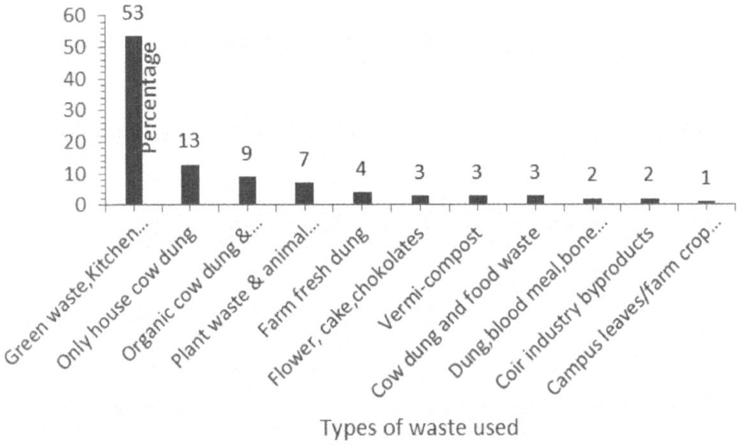

Fig. 2 Different types of feed-stock used by composting companies

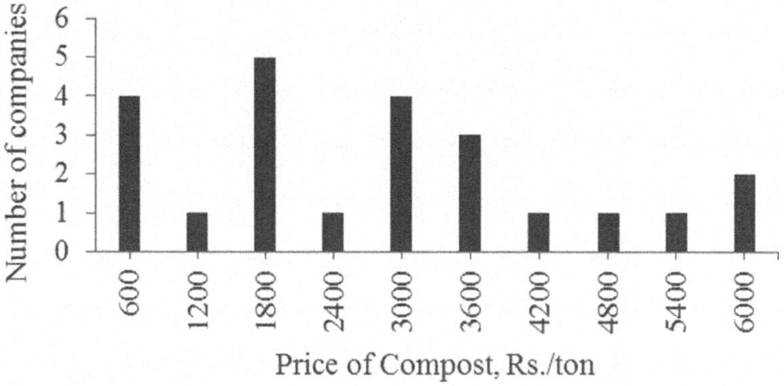

Fig. 3 Chart showing the prices of different composts

3.1.1 Price of Compost

Compost companies fix the compost prices based on demand. Among the 101 compost companies surveyed, only 23 companies announced their price on the Internet; it varied from Rs. 600/- per ton to Rs. 6,000/- per ton and the remaining compost companies mentioned 'optional' in their rate column. A histogram showing the number of companies and the prices of compost is shown in Fig. 3. It is important to note that the state and central governments are subsidizing composting companies with the intent to popularize composting by reducing its price. ULBs that produce compost from the organic fraction of MSW on a large scale often find it difficult to sell their compost to farmers at prices of Rs. 5/kg or even free of cost. There is a general perception that the quality of compost is highly variable or toxic and therefore, compost produced using MSW is not worth buying.

3.1.2 Particle Size Distribution of the Compost Samples

Most compost companies do not mention the size of the compost. Among the 101 compost companies surveyed, only 15 companies have reported the size of their compost particles on their websites. It varied from 2 to 7 mm and the remaining compost companies mentioned compost size as "normal", or "as per the requirements of the customer". The particle size distribution of compost provided by these companies is shown as a histogram in Fig. 4. Among the 15 companies that provided detailed size analysis data, 7 companies' composts were of 3 mm size, one company's compost had a maximum size of 7 mm size and only one had the minimum size of 2 mm. As per Indian standards, 90% of the compost material should be less than 4 mm in size. Therefore, only 8 companies met the standards for particle size.

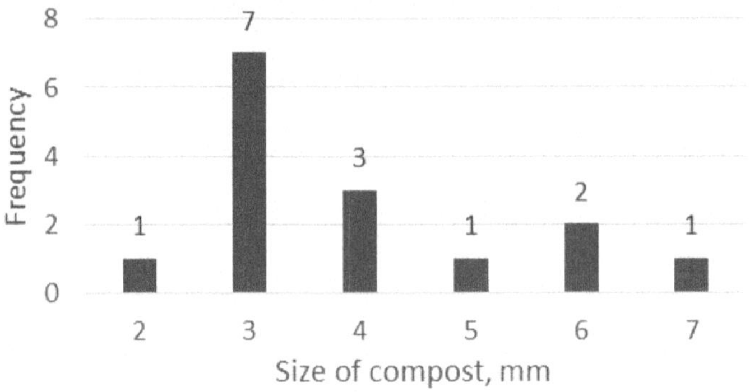

Fig. 4 Histogram showing particle sizes of different composts

3.2 *EDX Analyses*

It is necessary to determine the elements present in compost to evaluate the compost quality and applicability for agricultural purposes. Elemental analysis of essential nutrients like C, O, N, P, K, and S present in the compost samples was carried out for 11 samples and the per cent composition is depicted in Fig. 5. All compost samples in the present study have an abundant quantity of oxygen. The maximum quantity of oxygen was found in OC1 (46.33%), whereas the lowest quantity of oxygen was observed in DHP2 (33.16%). Essential nutrients like C, N, P, and K are in low quantities in most of the compost samples (Fig. 5). As all composts contain soil, large amounts of silicon were found in all compost samples (35.62% in OC 1 and 13.14% in NPC).

Various metals were found in the compost samples tested (Al, Ca, Mn, Fe, Mg, Cu, Zn, Ti, Pb, Hg, and Cr) but in less quantity compared to the previously mentioned

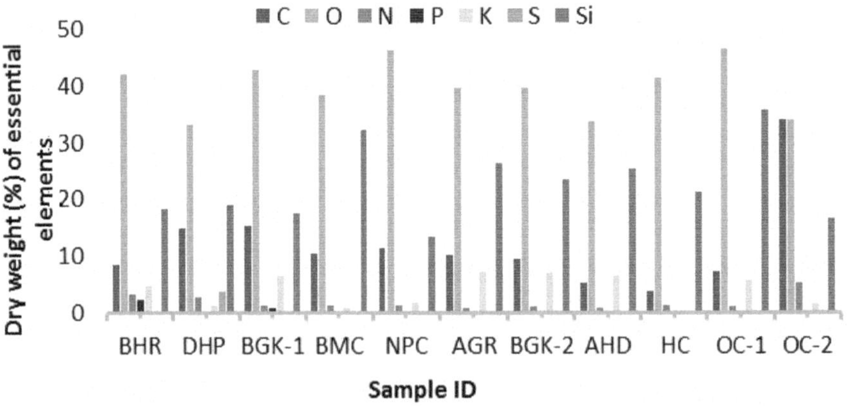

Fig. 5 Essential elements present in compost samples

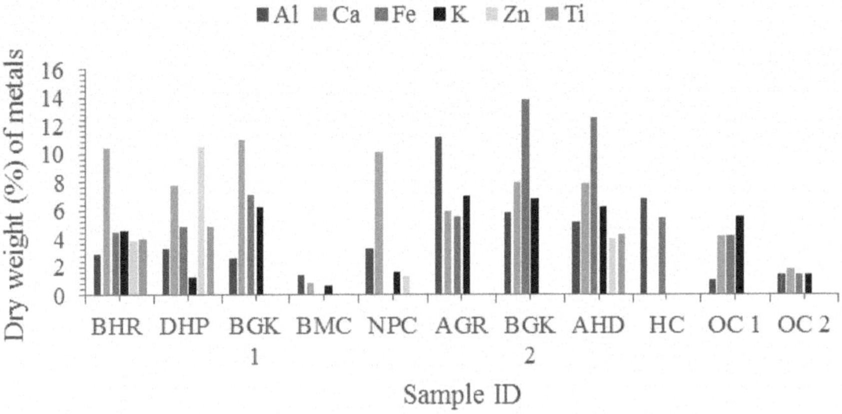

Fig. 6 Metals present in compost samples

elements. Metals like Al, Si, Fe, K, Zn, and Ti were found to be more than 2% by the total percentage weight of elements in compost samples (Fig. 6). Sample AGR had the highest percentage (11%) of aluminum (agriculture soil has 5.52% of Al) and BMC had the least (1%). Other metals like Fe and K were found in almost all samples whereas Zn and Ti were found in BHR 2, DHP 2, and AHD in small quantities. Many of these metals like Al, Ca, Fe, K, Zn, Ti, Mg and Mn are considered to be micro-nutrients which means they are useful and essential for plant growth in trace amounts.

Some toxic heavy metals were also found in the compost samples (As, Pb, Hg, Cd, Cr) but in lesser quantities than the major elements. These five metals have no known useful properties and are considered to be toxic at all concentrations. Arsenic in NPC was the highest at 0.38% of the dry weight of the sample, which is equivalent to 3800 ppm. The compost quality standard for As is 10 ppm (CPCB 2006). Two other samples: BMC (1400 ppm) and OC2 (200 ppm) were also in exceedance of the As standard (Fig. 7). Sample OC2 is in exceedance of compost standards for As, Pb, Hg, Cd, and Cr. These results prove that at least three compost samples were contaminated by toxic heavy metals.

Composts containing large amounts of heavy metals like Cr, Cd, Zn, Cu, Ni and Pb are a major concern.

EDX graphs of NPC and BMC samples are shown in Figs. 8 and 9, respectively.

SEM images of Neem Patta compost and Bone Meal compost are shown in Figs. 10 and 11. Both images show the granular nature of compost particles. The sizes of the grains vary from 2 to 4 microns and are likely to be silica nodules based on EDX analyses and other observations (Phutela & Sahni, 2013).

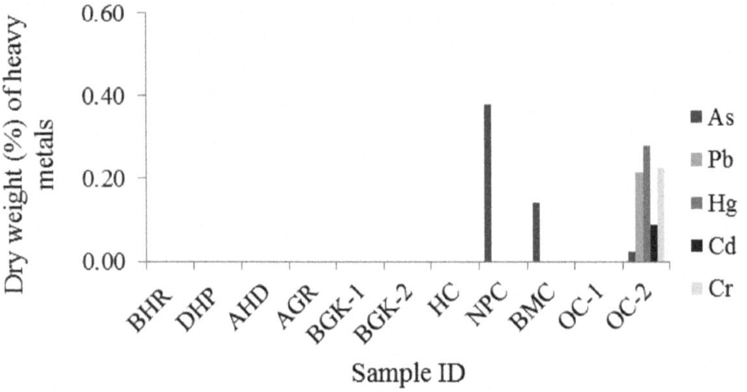

Fig. 7 Heavy metals present in compost samples

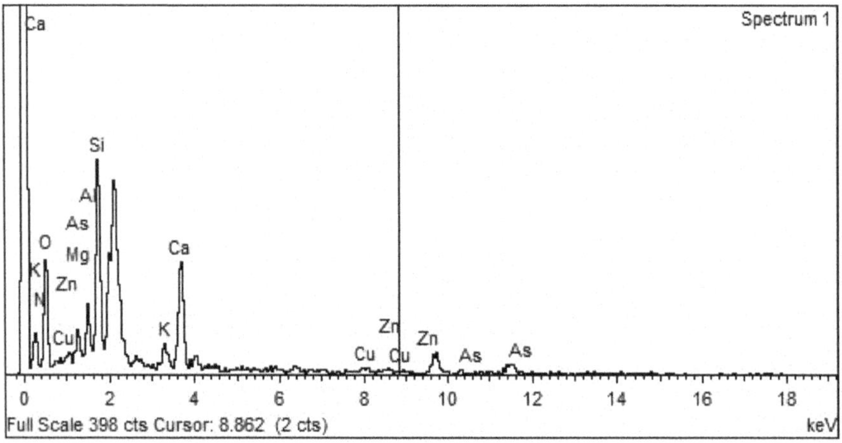

Fig. 8 EDX of NPC compost

4 Conclusions

Composting is one of the best methods of recycling organic materials. The objective of this study was to do a quantitative and qualitative analysis of composting companies and compost samples in India. Composting companies were surveyed based on their Internet websites and compost quality was determined by laboratory analysis. Eleven compost samples were collected and tested for their elemental composition using EDX analyses. The elemental analyses report shows that essential nutrients like N, P, and K are in less than required quantities which indicates poor compost quality. The major conclusions from this study were:

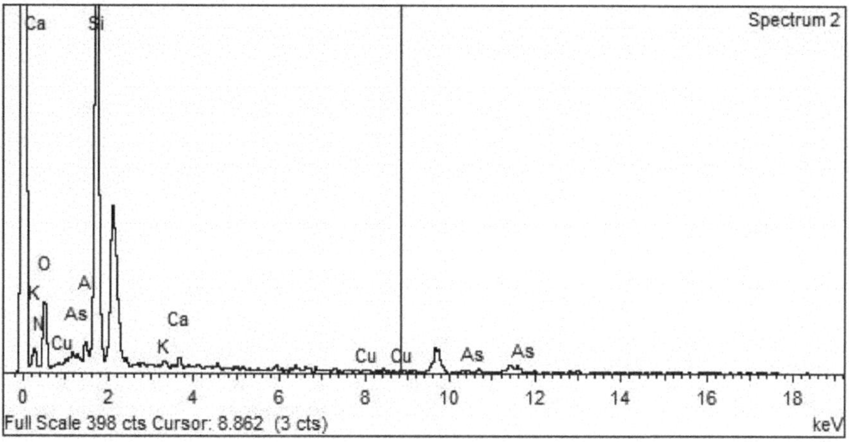

Fig. 9 EDX of BMC compost

Fig. 10 SEM image of Neem-Patta compost

- The largest numbers of compost companies (with Internet addresses) were in Karnataka.
- Most composting companies use green waste and kitchen waste for composting.
- Compost companies produce different types of compost such as commercial compost, vermicompost, farm/manure compost, mixed municipal compost, and forest/woodchip compost.
- Most of the composts sampled in this study are not harmful for use in agriculture; 8 out of 11 were non-toxic while the other 3 had toxic heavy metals in them. However, even the non-toxic compost samples were of poor quality due to their low nutrient levels.

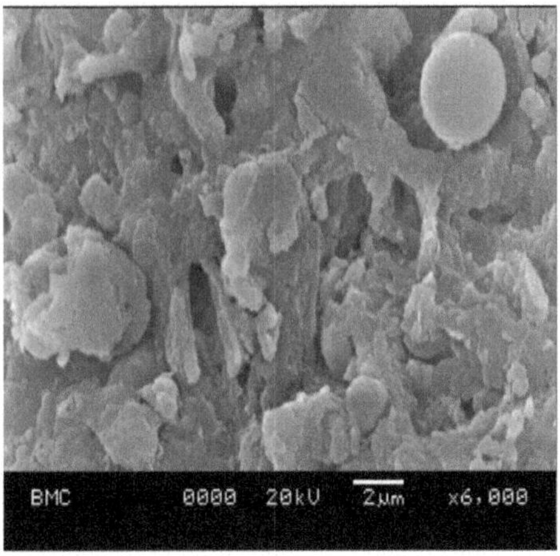

Fig. 11 SEM image of bone meal compost

References

Adani, F., Tambone, F. and A. Gotti (2004). Biostabilization of municipal solid waste. *Waste Management*, **24**: 775–783.

Bhatia, A., Madan, S., Jitendra, S., Muntjeer, A. and R. Pathania (2013). Diversity of bacterial isolates during full scale rotary drum composting. *Waste Management*, **33**: 1595–1601.

Chang, J.I. and Y.J. Chen (2010). Effects of bulking agents on food waste composting. *Bioresource Technology*, **101**: 5917–5924.

CPCB (2006). Characterization of MSW Compost and its application in agriculture. New Delhi, India. Downloaded from http://cpcbenvis.nic.in/scanned%20reports/CUPS%2059.pdf.

CPCB (2013). Status Report on Municipal Solid Waste Management. Central Pollution Control Board (Ministry of Environmental and Forests) Government of India Parivesh Bhawan, East Arjun Nagar.

Gaind, S. and A.C. Gaur (2000). Quality assessment of compost prepared from fly ash and crop residue. *Bioresource Technology*, **87**: 125–127.

Kalemelawa, F., Nishihara, E., Endo, T., Ahmad, Z., Yeasmin, R., Tenywa, M.M. and S. Yamamoto (2012). An evaluation of aerobic and anaerobic composting of banana peels treated with different inoculums for soil nutrient replenishment. *Bioresource Technology*, **126**: 375–382.

Kumar, M., Ou, Y.L. and J.G. Lin (2010). Co-composting of green waste and food waste at low C/N ratio. *Waste Management*, **30**: 602–609.

Manios, T. (2004). The composting potential of different organic solid wastes: experience from the island of Crete. *Environment International*, **29**: 1079–1089.

Phutela, U.G. and N. Sahni (2013). Microscopic structural changes in paddy straw pretreated with *Trichoderma reesei* MTCC 164 and *Coriolus versicolor* MTCC 138. *Indian Journal of Microbiology*, **53(June)**: 227–231. https://doi.org/10.1007/s12088-012-0321-8.

Ribeiro, H., Vasconcelos, E. and J.Q. Dos Santos (2000). Fertilisation of potted geranium with municipal solid waste compost. *Bioresource Technology*, **73**: 247–249.

Rosen, C., Halbach, T. and B. Swanson (1993). Horticultural uses of municipal solid waste composts. *HorTechnology*, **3**: 167–173.

Chapter 4
Treatment Options for Efficient Municipal Solid Waste Management

B. R. Hiremath, Anuja Joseph, and Sudha Goel

1 Introduction

Municipal solid waste (MSW) quantities are growing rapidly all over the world and more so in countries where urban populations and economies are growing together. India is a good example of this phenomenon. This growth is driven mainly by two factors: rapid increase in urban population and increase in per capita resource consumption which in turn leads to greater per capita waste generation rates. Urban Local Bodies (ULBs) in India are struggling to carry out their responsibilities of collection, transportation, treatment and disposal of MSW appropriately. Lack of sufficient information, experience, technical know-how and financial resources are some of the major barriers to achieve efficient MSW management. A brief literature review of MSW generation rates and composition along with treatment options used in India is provided in this chapter.

B. R. Hiremath
Civil Engineering Department, Basaveshwar Engineering College, Bagalkot, Karnataka 587 102, India

A. Joseph
School of Environmental Science & Engineering, Indian Institute of Technology Kharagpur, Kharagpur 721 302, India

S. Goel (✉)
Civil Engineering Department and School of Environmental Science and Engineering, Indian Institute of Technology Kharagpur, Kharagpur 721 302, India
e-mail: sudhagoel@civil.iitkgp.ac.in

S. Goel (ed.), *Advances in Solid and Hazardous Waste Management*,
https://doi.org/10.1007/978-3-031-49144-3_4

2 Bibliometric Analysis

2.1 Keyword Analysis and Network Mapping

In order to understand some of the most important aspects of MSW management over the last decade, a bibliometric analysis was conducted on 07 June 2021. Since the year 2011, a total of 3,784 scientific documents on municipal solid waste management were retrieved from the Web of Science database. The data were further analyzed using data visualization software—VOSviewer. Figure 1 shows the network mapping of the keywords occurring in the scientific documents. The size of the keywords varies with respect to the number of occurrences in the entire database as shown in Table 1. The network mapping, therefore, indicates the current emphasis on processes or methods like life cycle assessment, anaerobic digestion, landfill, recycling, circular economy, and composting.

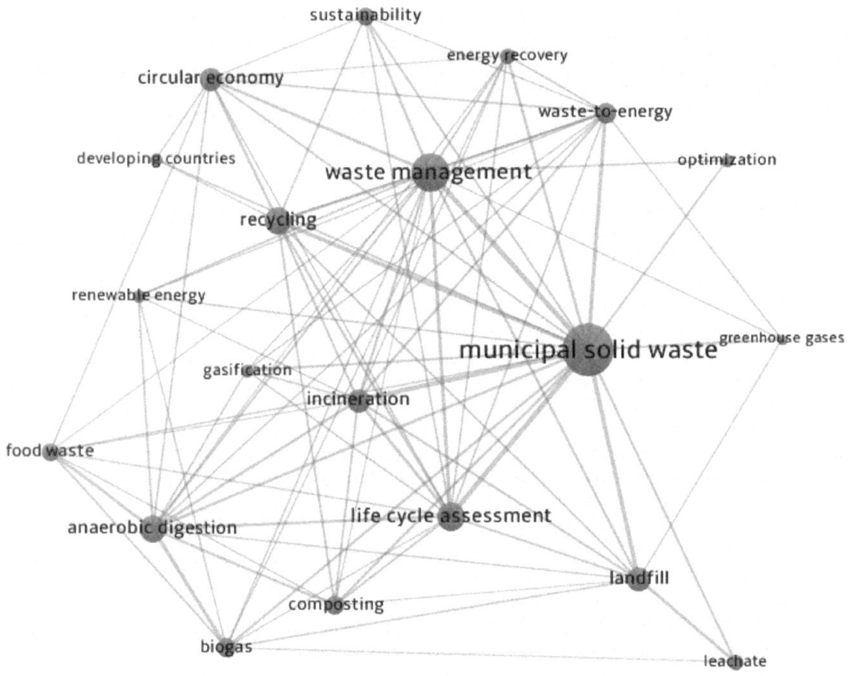

Fig. 1 Bibliometric analysis (2011–2021): network visualization of co-occurrence of keywords

Table 1 Occurrences and link strength of the keywords mapped in Fig. 1

S. No	Keywords	Occurrences	Total link strength
1	Municipal solid waste	976	
2	Waste management	575	
3	Life cycle assessment	213	151
4	Landfill	170	125
5	Anaerobic digestion	150	209
6	Recycling	150	211
7	Waste-to-energy	144	110
8	Circular economy	115	136
9	Incineration	112	222
10	Biogas	85	131
11	Composting	84	116
12	Sustainability	74	88
13	Food waste	71	82
14	Energy recovery	56	93
15	Leachate	56	61
16	Developing countries	43	55
17	Optimisation	41	41
18	Renewable energy	37	68
19	Gasification	36	60
20	Greenhouse gases	26	43

2.2 *Bibliometric Analysis of Treatment Options Globally and in India*

Bibliometric analysis of treatment methods used for municipal solid waste on a global scale was done using Scopus on 7 June 2021. A search was done using "Municipal solid waste treatment methods" in the article title, abstract or keywords. Only journal articles, reviews, books or book chapters in the English language were chosen for this bibliometric analysis. Conference articles were not included since they are often not accessible through the Internet.

Only 2512 publications were found and the first publication on this topic was in 1972 as shown in Fig. 2. The number of publications each year has grown exponentially after 1993. The top 10 journals with publications in this area are shown in Fig. 3 and Waste Management, the number one journal, has published 11.7% of the total publications on this topic. The top 10 countries with publications on this topic are shown in Fig. 4.

The global list of publications was then limited to 'India' as an additional keyword in Scopus on 7 June 2021 and a total of 357 publications were found of which 173 publications were from India as shown in Fig. 5. Other countries with publications

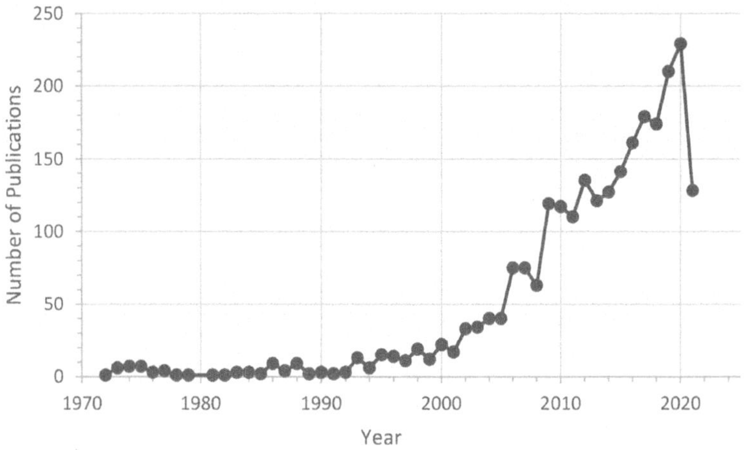

Fig. 2 Number of publications per year on the topic

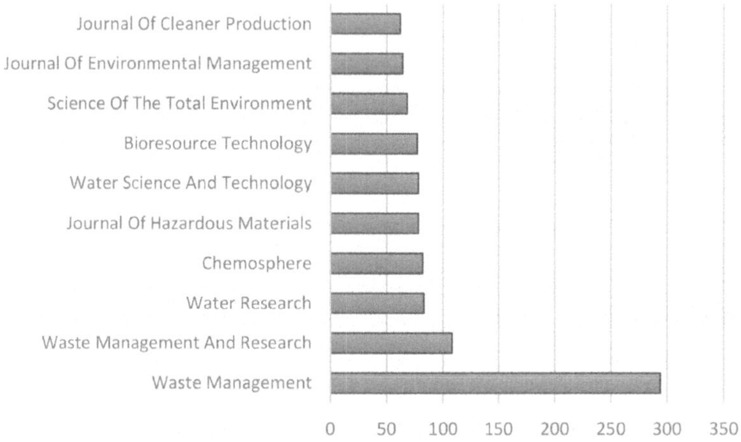

Fig. 3 Top 10 journals with publications on this topic

using the keyword 'India' are shown in Fig. 5. The number of publications per year is shown in Fig. 6 and it is evident that based on publications, research activity on this topic started in 1987 in India and has been growing steadily only after 2002. Of the 357 publications, 288 were articles, 48 were reviews, 19 were book chapters and 2 were books.

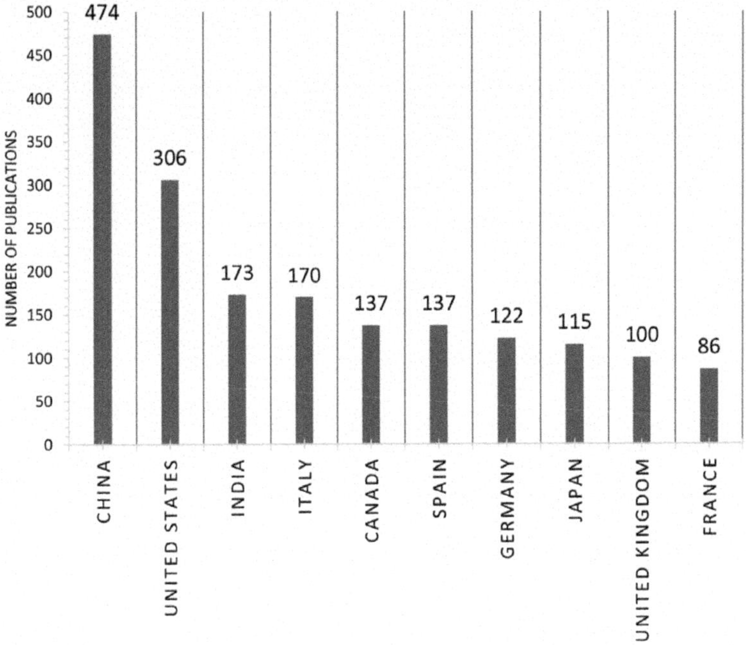

Fig. 4 Top 10 countries with publications on this topic

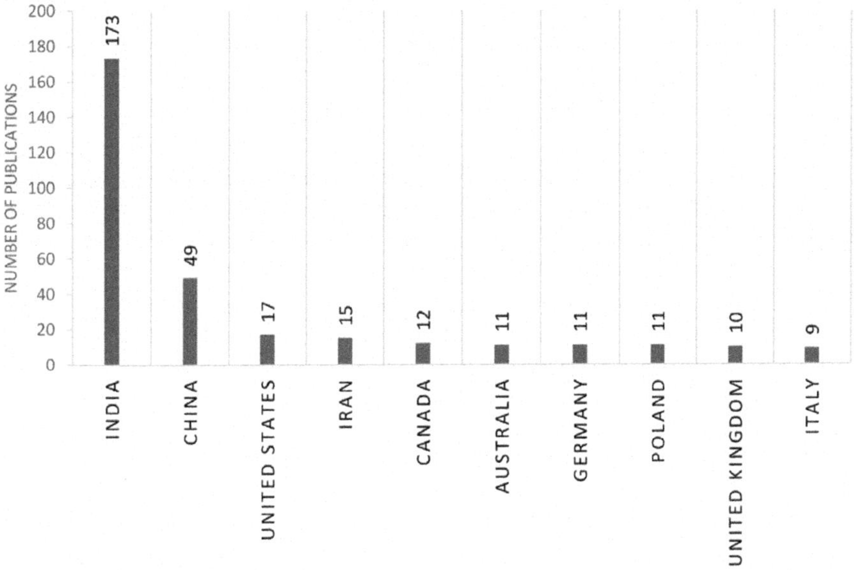

Fig. 5 Number of publications from different countries with India as an additional 'keyword'

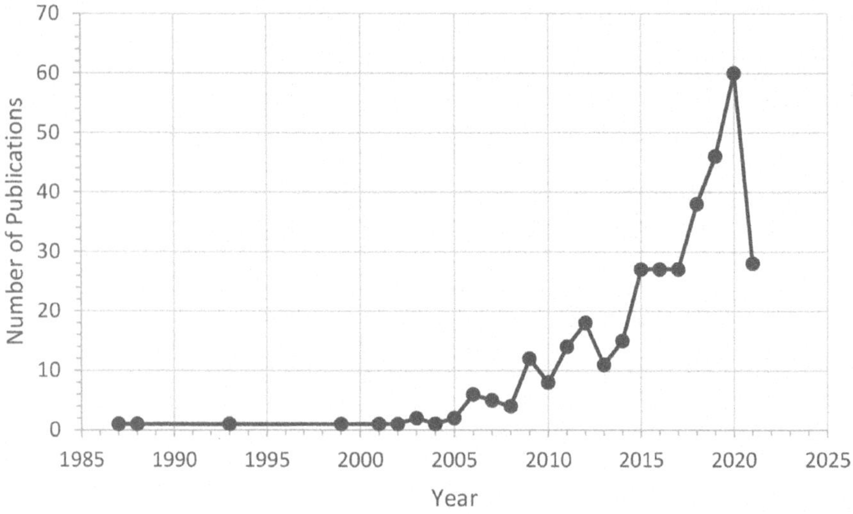

Fig. 6 Number of publications regarding treatment methods used for MSW in India

3 MSW Generation Rates

Proper solid waste management requires complete information regarding quantities of waste generated and the composition of the waste. A review of the factors affecting the quantities of MSW generated shows that population size, income level and location-specific factors are the most important (Goel et al. 2017). Total MSW generation rates are directly proportionate to population, while per capita MSW generation rates are directly proportionate to economic levels and the size of the household. Other factors that impact MSW generation rates are seasonal variations, geographic or climatic factors, government policies and legislation. In a related study conducted in China, economy and urban development were the most important factors in determining MSW generation rates (Liu and Wu 2011).

Solid waste generation rates have been growing rapidly in most countries and more so in developing countries with growing economies. Global solid waste generation in the year 2016 was estimated to be 2.01 billion tons and is expected to reach 3.4 billion tons by 2050 (Kaza et al. 2018). Waste generation rates are related to income levels. High income countries were responsible for 34% of the global waste generated, lower middle- and higher-middle-income countries contributed 29% and 32%, respectively, while lower income countries contributed only 5% to the total global waste generated. The global average per capita generation rate of solid waste was 0.74 kg/capita-day and ranges from 0.11 to 4.54 kg/capita-day. MSW generation is related to urban areas only and the total solid waste generated in different countries showed a clear correlation with urbanization rates.

Urban MSW generation in India in 2018–2019 was 152,076.7 metric tons/day of which 149,748.6 tons/day was collected, 55,759.6 tons/day was treated and 50,161.33

tons/day was landfilled (CPCB 2019). The remaining waste is generally dumped on open ground. These data show that MSW collection was 98.5, 36.7% of the waste generated was treated and 33% of the waste was landfilled. No information is available regarding the amount of waste that is recycled. The recent estimate of the average per capita MSW generation rate in Indian cities is 0.6 kg/person-year.

4 Composition of Municipal Solid Waste

The average composition of solid waste depends mainly on the income levels of countries (Kaza et al. 2018). The global average waste composition is shown in Table 2 and significant differences in waste composition are evident as the income levels of countries increase.

The composition of waste can change over time, especially in a growing economy like India. These changes are expected based on changes in resource consumption. As shown in Table 3, the fraction of paper and plastic waste has increased during the period from 1996 to 2011, while the inert fraction of MSW has decreased. As the economy improves, the usage of paper and plastic increases, resulting in more waste of these components. The inert fraction of MSW has decreased in most towns and cities due to an increase in paved areas which leads to less soil, dirt and dust in MSW and conversion from the use of coal to gas for cooking in most households. Use of cooking gas results in no ash while coal-based cooking contributes a large amount of ash to MSW (IBRD 1999).

A study conducted in Dhanbad, India found that lower socio-economic groups generated approximately 30% ash residue due to coal usage and around 20% food waste as the major content in the solid waste generated (Khan et al. 2016).

Table 2 Global solid waste composition for countries with varying income levels (Kaza et al. 2018)

Income level	Food and green	Glass	Metal	Paper and cardboard	Rubber and leather	Plastic	Wood	Other
Global average	44	5	4	17	2	12	2	14
Lower	56	1	2	7	<1	6.4	<1	27
Lower middle	53	3	2	12.5	<1	11	1	17
Upper middle	54	4	2	12	1	11	1	15
Higher	32	5	6	25	4	13	4	11

Table 3 Change in composition of municipal solid waste in India with time (in % on a dry weight basis) (Planning Commission Report as cited in Joshi and Ahmed (2016))

Year	Biodegradables	Paper	Plastic/rubber	Metal	Glass	Rags	Others	Inert
1996	42.21	3.63	0.60	0.49	0.60	–	7.94	45.13
2005	47.43	8.13	9.22	0.50	1.01	4.49	4.06	25.16
2011	42.51	9.63	10.11	0.63	0.96	–	19.16	17.00

5 Municipal Solid Waste Treatment Options

MSW can be treated in several ways prior to final disposal by incineration or landfilling. In India, composting (50%), anaerobic digestion (30%), pelletization (10%), and sanitary landfilling (10%) were the most commonly adopted technologies for municipal solid waste management based on a government report published in 1993 (Palanichamy et al. 2002; Singh et al. 2011a, b).

More recent data show that most ULBs have opted for composting as a treatment process (CPCB 2018). Other options being used by ULBs are vermicomposting, biogas generation by anaerobic digestion and refuse-derived fuel (RdF) generation or pelletization. The number of plants set up by state governments or ULBs is shown in Fig. 7 along with the number of operational plants in the country.

The success of any treatment process is dependent on the nature of the waste fed to the process. This necessitates the segregation of the waste components, ideally at source into three different fractions: wet waste (food, kitchen, garden, and yard waste) and dry waste (non-biodegradable waste) and recyclables. In most households in India, waste is not segregated into wet waste (food, kitchen, garden, and yard waste) and dry waste (non-biodegradable waste). If this level of source segregation is not practiced in every household along with door-to-door collection, then manual or mechanical segregation becomes necessary at the treatment plant.

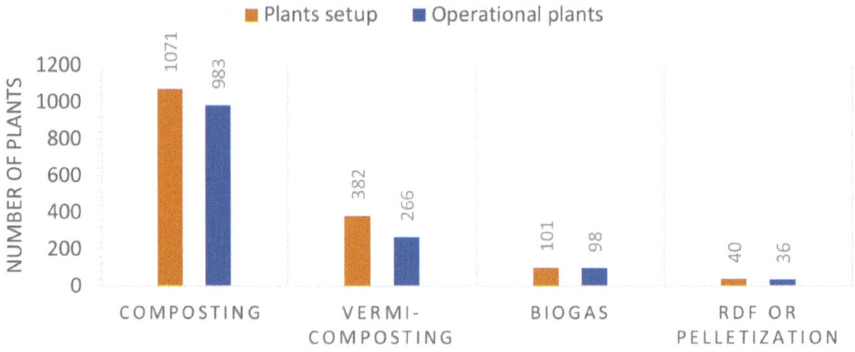

Fig. 7 Number of plants setup in India with different treatment processes and number of operational plants (based on CPCB 2018 data)

A study conducted in Delhi, India showed that 60% of the residents were unaware of biodegradable and non-biodegradable waste whereas only 2% of the population regularly segregated their waste (Mukherji et al. 2016). According to the survey, approximately 50% of respondents claimed that the waste collector would mix the already segregated wastes while 97% of the respondents gave their recyclables to waste buyers who come to their doors. These results suggest that greater awareness and education are required for implementing source segregation and ULBs have to provide appropriately designed collection vehicles for the different segregated and recyclable fractions of MSW.

Many municipalities hire private firms for managing waste collection and treatment, whereas some employ non-government organizations for supervising and collecting segregated wastes. Community bins are common in India where the entire neighborhood dumps wastes into these bins without segregation, while the government is now trying to achieve 100% door-to-door collection of waste in all urban centers.

5.1 Sanitary Landfill

Disposing of municipal solid wastes on open land (private or publicly-owned) without any treatment is the final fate for most MSW in low and middle income countries all around the world (Kaza et al. 2018). Almost all of the MSW (93%) was dumped on open land in low income countries and 30–66% of the waste was in middle-income countries. In contrast, only 2% of the total MSW was dumped on open land in high-income countries. Loss of a valuable resource such as land in middle-income countries like India is a matter of concern for the government and the public. Historical satellite images have been used to measure the land area occupied by these open dumpsites (Mahmood et al. 2019; Singhal and Goel 2019).

The sanitary landfill was designed by the American Society for Civil Engineers (ASCE) in the year 1959 to combat open dumping. In general, residuals from recovery and recycling activities and combustion ash are placed in landfills. The residues consist mostly of plastic, newspaper, broken glass, dirt and ash. The sanitary landfill system has two outputs: liquid (leachate) as well as gases. The discharge of the leachate and gas needs to be continuously regulated, contained and collected based on the rules and regulations created by the concerned authorities. Leachate is treated before discharge to sewers or a nearby surface water body (Youcai, 2018).

Sanitary landfills must be designed properly and a recently constructed landfill at Baidyabatinear Kolkata, India is shown in Fig. 8. These landfills are designed with liners as per regulations, leachate control systems, collection of landfill gases, and other control systems. Proper waste disposal techniques have to be followed to operate such a landfill effectively. Therefore, the cost of constructing and maintaining a sanitary landfill site increases drastically. In developing countries like India, where land is a valuable asset as well as a vital resource, construction and maintenance of sanitary landfills is increasingly difficult. The current regulations require that only

Fig. 8 Sanitary landfill at Baidyabati near Kolkata. It is operated in phases. The area on the left is completely filled while the area on the right is ready to be used. The liner, leachate and gas collection systems are already in place in the area on the right

non-usable, non-recyclable, and non-biodegradable waste should be placed in a sanitary landfill. The landfill shown in Fig. 8 was built by a consortium of 6 municipalities in the suburban areas around Kolkata because none of the municipalities could have constructed or operated properly designed facilities on their own.

5.2 *Composting*

Composting is a traditional method followed in India and other countries for retaining humus in the soil and maintaining soil fertility. Composting requires very little financial investment, low technical skills and is not labour intensive, unlike other treatment processes. Composting is useful for treating wastes like agricultural waste, horticultural waste, garden and kitchen wastes, and other wastes rich in organic content. If compost is generated from agricultural residue, it is termed farm compost while compost generated from municipal solid waste (MSW) is termed city compost. City compost may be contaminated with heavy metals, plastics and other undesirable materials due to the nature of the feed waste material making source segregation essential for city composting.

There are several methods for composting waste materials in India.[1] Some of the most common methods for farm composting are the Indore Method, the Nadep Method,[2] Coimbatore Method and the Bangalore Method. The oldest documented method of composting is the Indore process which was formulated and implemented in Indore by Sir Albert Howard in 1933 (Rao 2015).

Composting is the conversion of organic components of residential wastes into humus like material. This final product can then be used as a soil conditioner or supplemental fertilizer. Depending on the frequency of turning or mixing of the waste, composting can be aerobic or anaerobic. Various organisms are capable of composting waste but conventional composting is carried out mainly by microbes like bacteria and fungi. Heat is generated during the metabolic processes and some of the carbon in the waste is mineralized to carbon dioxide. This results in a compost that is rich in nitrogen and has a lower carbon content than the feed material (Azim et al. 2018).

Diversion of the organic fractions of waste to the composting process can reduce the burden on sanitary landfills, reduce pollution of air, groundwater, and soil and reduce emissions of greenhouse gases like methane and carbon dioxide (Pergola et al. 2018). Composting can be carried out in windrows, static aerated piles, trenches or pits, or reactors (Kumar 2011).

The Indore method uses waste mainly from animal feed lots and farms. A trench was excavated with dimensions of 30 feet × 14 feet × 2 feet deep.[3] If the pit is deeper than $2^{'}$, the waste can become anaerobic. The sides were sloped and the depth of the pit was 2 feet so that aeration could be maintained. Animal urine and dung were mixed with wood ash, soil, earth, straw (crushed agricultural waste), bacterial and fungal inoculum and vegetable waste. The pit was charged for 6 days and then watered or turned at frequent intervals for a 90-day period as shown in Table 4. A maximum temperature of 65 °C was achieved at 12 h and moisture content was maintained between 40 and 60% by water at frequent intervals throughout the composting period. Excess moisture can lead to nitrogen losses, anaerobic conditions and loss of heat or decreased maximum temperature due to evaporation of moisture.

5.3 Vermi-Composting

Vermicomposting is the degradation of organic matter by earthworms and microorganisms. Earthworms are known to improve soil conditions by tilling the soil through their movement. They also ingest waste materials which are broken down in their guts by their coeliac fluids and microorganisms (Singh et al. 2011a, b). The excreta from their digestion process is known as vermicasts or castings. These castings are

[1] https://www.slideshare.net/praveenentomo/composting-methods-and-techniques-praveenbpati 1210514.

[2] http://ecoursesonline.iasri.res.in/mod/page/view.php?id=149590.

[3] http://journeytoforever.org/farm_library/HowardWPA/WPA4.html.

et al. 2018). An external source of energy is required to drive the pyrolysis process since the conversion reactions are endothermic in nature.

One of the main disadvantages of the pyrolysis process is the complex nature of the product stream when compared to other alternative treatments. The product gases have high concentrations of carbon monoxide. Therefore, immediate treatment is required before venting the gas directly into the atmosphere (Dreyer et al. 2015).

Plasma pyrolysis is utilized for destroying plastic waste in the absence of oxygen. A plasma torch is used for the destruction of plastic waste at an extreme temperature of 1000 °C (CPCB 2017). Pollutants like carbon monoxide, dioxins, furans, nitrogen dioxide and particulate matter are liberated into the atmosphere. Therefore, the gases have to be treated before being released into the atmosphere.

5.8 Refuse-Derived Fuel (RdF)

Refuse derived fuel (RdF) is a type of fuel extracted by separating, drying and extruding the organic fraction of waste from mixed or segregated MSW and converting it into pellets or briquettes. It can be mixed with biomass waste, agricultural waste and sewage sludge. Refuse-derived fuel (RDF) has high calorific value and can be used as a co-fuel in power plants, in cement industries, and as alternatives for industrial boilers (Ouda et al. 2016; Dong et al. 2018; Hossain et al. 2019; Khan et al. 2016). RdF can reduce the amount of waste going to landfills, reduce emissions of greenhouse gases, and avoid uncontrolled usage of fossil fuels.

The organic fractions of MSW which have low moisture content (15–20%) and high calorific value (2200–3000 kcal/kg) can be used as raw material for refuse-derived fuel. The treatment methods used to process MSW prior to generating RdF are size reduction, drying, sorting, screening, separation, and pelletization (Aluri et al. 2018).

5.9 Mechanical-Biological Treatment (MBT)

Mixed MSW cannot be treated by a single treatment process and requires several treatment methods to be integrated into a single facility for effective SWM. The idea of developing a single treatment facility with mechanical and biological treatment (MBT) became popular in Europe from 1990 onwards mainly due to a ban in 1999 on disposing off biodegradable materials in sanitary landfills (Montejo et al. 2013; Fei et al. 2018). MSW rules in India became effective in 2000 and required that MSW should be treated prior to disposal and that anything biodegradable should not be disposed off in sanitary landfills. Since segregation of waste at source is still not common in India, MBT facilities for treating mixed MSW are becoming the norm in most towns and cities. The mixed waste is segregated in these facilities using various mechanical processes such as sieving, grinding, magnetic separation, air classification, spin drying or solar drying and recyclables are collected from the mechanical part of the MBT. The organic fraction is then taken for composting or anaerobic digestion depending on the characteristics of this waste fraction.

As the quantities of MSW are growing exponentially all over the world, it is becoming more important to find the right treatment options for handling these wastes and reducing the burden on land in the form of sanitary landfills. Research and development activity in the area of 'MSW treatment methods' has increased tremendously after 1993 at a global scale and received attention in India only after 2002–2003. Choosing and designing an appropriate MSW treatment facility including an MBT requires knowledge of generation rates and characteristics of the mixed MSW. An efficient integrated waste treatment facility can only be designed if these data are collected on a regular basis and are accurate. The choice of treatment methods in an MBT or other treatment facility depends on the nature of MSW collected, i.e., it should be either mixed or segregated at the source. At present, MBT plants are the only solution to the mixed waste problem. However, it is possible that as citizens and ULBs focus on source segregation, MBTs may be replaced by better and more specialized treatment processes.

References

Aluri, S., et al. (2018). Pyrolysis and gasification studies of model refuse derived fuel (RDF) using thermogravimetric analysis. *Fuel Processing Technology*, **179**: 154–166. https://doi.org/10.1016/J.FUPROC.2018.06.010.

Azim, K., et al. (2018). Composting parameters and compost quality: A literature review. *Organic Agriculture*, **8**(**2**): 141–158. https://doi.org/10.1007/s13165-017-0180-z.

Bufoni, A.L. (2017). The change from aerobic to anaerobic landfilling and increase of correspondent emissions and expenses. *In:* Sengupta, D. and S. Agrahari (eds) Modelling Trends in Solid and Hazardous Waste Management. 1st ed. Singapore: Springer Nature, pp. 141–158.

CPCB (2017). *Annual Review Report: 2015–16*. Delhi. Available at: www.cpcb.nic.in.

CPCB (2018). *Consolidated annual report (for the year 2016–2017) on implementation of solid waste management rules, 2016*. Delhi. Available at: www.cpcb.nic.in.

CPCB (2019), Annual Report 2018-19 on Implementation of Solid Waste Management Rules, Available at https://www.cpcb.nic.in

Dong, J., et al. (2018). Life cycle assessment of pyrolysis, gasification and incineration waste-to-energy technologies: Theoretical analysis and case study of commercial plants. *Science of The Total Environment*, **626**: 744–753. https://doi.org/10.1016/J.SCITOTENV.2018.01.151.

Dreyer, J.A.H., et al. (2015). Preferential oxidation of carbon monoxide over Pt–FeO_x/CeO_2 synthesized by two-nozzle flame spray pyrolysis. *Journal of Catalysis*, **329**: 248–261. https://doi.org/10.1016/J.JCAT.2015.05.003.

Fei, F., et al. (2018). Mechanical biological treatment of municipal solid waste: Energy efficiency, environmental impact and economic feasibility analysis. *Journal of Cleaner Production*, **178**: 731–739. https://doi.org/10.1016/j.jclepro.2018.01.060.

Goel, S., et al. (2017). Forecasting Solid Waste Generation Rates. *In:* Sengupta, D. and S. Agrahari (eds.). Modelling Trends in Solid and Hazardous Waste Management. Springer, Singapore. https://doi.org/10.1007/978-981-10-2410-8_3.

Gómez, N., et al. (2018). Effect of temperature on product performance of a high ash biomass during fast pyrolysis and its bio-oil storage evaluation. *Fuel Processing Technology*, **172**: 97–105. https://doi.org/10.1016/J.FUPROC.2017.11.021.

Gupta, C., et al. (2019). Role of Vermicomposting in Agricultural Waste Management. *In:* Shah, S., Venkatramanan, V. and R. Prasad (eds). *Sustainable Green Technologies for Environmental Management*. Singapore: Springer Singapore, pp. 283–295. https://doi.org/10.1007/978-981-13-2772-8_15.

Hossain, M.U., et al. (2019). Techno-environmental feasibility of wood waste derived fuel for cement production. *Journal of Cleaner Production*, **230:** 663–671. https://doi.org/10.1016/J.JCLEPRO.2019.05.132.

Huber, F., et al. (2016). Thermal co-treatment of combustible hazardous waste and waste incineration fly ash in a rotary kiln. *Waste Management*, **58**: 181–190. https://doi.org/10.1016/j.wasman.2016.09.013.

IBRD, W.B. (1999). *What a Waste: Solid Waste Management in Asia*. Washington, DC.

Joshi, R. and S. Ahmed (2016). Status and challenges of municipal solid waste management in India : A review. *Cogent Environmental Science*, Cogent, **28(1)**: 1–18. https://doi.org/10.1080/23311843.2016.1139434.

Kaza, S., Yao, L., Bhada-Tata, P. and F. Van Woerden (2018). *What a Waste 2.0: A Global Snapshot of Solid Waste Management to 2050*. Washington DC, USA: International Bank for Reconstruction and Development, The World Bank, Washington, DC.

Khan, D., Kumar, A. and S.R. Samadder (2016) Impact of socioeconomic status on municipal solid waste generation rate. *Waste Management*, **49**: 15–25. https://doi.org/10.1016/j.wasman.2016.01.019.

Kumar, S. (2011). Composting of municipal solid waste. *Critical Reviews in Biotechnology*, **31(2)**: 112–136. https://doi.org/10.3109/07388551.2010.492207.

Liu, C. and X. Wu (2011). Factors influencing municipal solid waste generation in China: A multiple statistical analysis study. *Waste Management & Research*, **29(4)**: 371–378. https://doi.org/10.1177/0734242X10380114.

Mahmood, K., et al. (2019). A comparison of satellite-based indices for hazard assessment of MSW open dumps using spatial analysis. *Waste Management and Research*, **37(3)**: 219–236. https://doi.org/10.1177/0734242X18815963.

Montejo, C., et al. (2013). Mechanical-biological treatment: Performance and potentials. An LCA of 8 MBT plants including waste characterization. *Journal of Environmental Management*, **128**: 661–673. https://doi.org/10.1016/j.jenvman.2013.05.063.

Moustakas, K. and M. Loizidou (2010). *Solid Waste Management through the Application of Thermal Methods, Waste Management*. *In:* Kumar, S. (ed.). Waste Management, Chapter 6, InTechOpen: London, UK. doi: https://doi.org/10.5772/8464.

Mukherji, S.B., et al. (2016). Resident knowledge and willingness to engage in waste management in Delhi, India. *Sustainability (Switzerland)*, **8(10)**: 1–14. https://doi.org/10.3390/su8101065.

Neubacher, F.P. and Q. Huang (2019). WTE: Fluidized Bed Technology. *In:* Themelis, N.J. and A.C. Bourtsalas (Thanos) (eds). Recovery of Materials and Energy from Urban Wastes: A Volume in the Encyclopedia of Sustainability Science and Technology, Second Edition. New York, NY: Springer New York, pp. 375–391. https://doi.org/10.1007/978-1-4939-7850-2_405.

Ngamsang, T. and M. Yuttitham (2019) Vulnerability assessment of areas allocated for municipal solid waste disposal systems: A case study of sanitary landfill and incineration. *Environmental Science and Pollution Research*, **26**: 27239–27258.

Ouda, O.K.M., et al. (2016). Waste to energy potential: A case study of Saudi Arabia. *Renewable and Sustainable Energy Reviews*, **61**: 328–340. doi: https://doi.org/10.1016/J.RSER.2016.04.005.

Palanichamy, C., Babu, N.S. and C. Nadarajan (2002). Municipal solid waste fueled power generation for India. *IEEE Transactions on Energy Conversion*, **17(4)**: 556–563.

Pergola, M., et al. (2018). Composting: The way for a sustainable agriculture. *Applied Soil Ecology*, **123**: 744–750. https://doi.org/10.1016/J.APSOIL.2017.10.016.

Rajesh, C., et al. (2019). Sustainable waste management through waste to energy technologies in India-Opportunities and environmental impacts. *International Journal of Renewable Energy Research,* **9**(**1**): 309–342.

Rao, D. (2015). *History of Composting*. https://doi.org/10.13140/RG.2.1.1408.8402.

Sharma, K. and V.K. Garg (2019). Vermicomposting of Waste: A Zero-Waste Approach for Waste Management. *Sustainable Resource Recovery and Zero Waste Approaches*. Elsevier, pp. 133–164. https://doi.org/10.1016/B978-0-444-64200-4.00010-4.

Singh, N. (2018) *Swachh Bharat Mission (Urban): Rethinking waste valorization in fast growing middle-income cities*. New Delhi.

Singh, R.P., et al. (2011) An overview for exploring the possibilities of energy generation from municipal solid waste (MSW) in Indian scenario. *Renewable and Sustainable Energy Reviews*, **15**(**9**): 4797–4808. https://doi.org/10.1016/j.rser.2011.07.071.

Singh, R.P. et al. (2011). Management of urban solid waste: Vermicomposting a sustainable option. *Resources, Conservation and Recycling*, **55**(**7**): 719–729. https://doi.org/10.1016/j.resconrec.2011.02.005

Singhal, A. and S. Goel (2019). An integrated solid waste management (ISWM) plan using Google Earth and Linear Programming: A case study of Kharagpur city, West Bengal. *In:* Sengupta, D., Dubey, B.K. and S. Goel (eds). Treatment and Disposal of Solid and Hazardous Wastes. 1st edn. New Delhi, pp. 169–192.

Sinha, R.K., et al. (2002). Vermiculture and waste management: Study of action of earthworms. *The Environmentalist*, **22**(**3**): 261–268.

Tian, H., et al. (2017). Acclimation to extremely high ammonia levels during continuous biomethanation process. *In:* 15th IWA World Conference on Anaerobic Digestion Conference. International Water Association: Beijing, China.

Youcai, Z. (2018). Leachate Treatment Engineering Processes. *Pollution Control Technology for Leachate from Municipal Solid Waste*. Butterworth-Heinemann, pp. 361–522. https://doi.org/10.1016/B978-0-12-815813-5.00005-X.

Chapter 5
Crop Residue Management Techniques for North-Western India: Evaluating Alternatives to Residue Burning

Nawazish Parwez, Bishwatma Biswas, and Sudha Goel

1 Introduction

Agriculture and its allied sectors significantly contribute to the Indian economy, employment and the environment. According to World Bank data (accessed in 2021), in 1960, agriculture, forestry and fishing together accounted for 41% of the country's GDP which decreased to 18.1% in 2020. More than 50% of the workforce in the country is associated with agriculture and allied activities making this sector the largest employment provider in India, especially in its vast rural regions (Bhatt et al. 2021). The area under agriculture accounts for 50.4–53.7% of the total area of the country (Wikipedia 2021).

In the Indo-Gangetic region which includes the northern parts of India and Pakistan, southern parts of Nepal and China and almost all of Bangladesh, the 'rice-wheat cropping system (RWCS)' is used. Cultivation is done over an area of more than 26 million ha and is responsible for staple food production for 1/5th of the world's population (Saharawat et al. 2010). To ensure food security in South Asia, wheat and rice production should increase by 1.7% and 1.1% per year, respectively, over the next four decades (Gathala et al. 2013). Although RWCS plays a significant role in agricultural activities, the sustainability of the system has become a major concern as it is linked to groundwater depletion, soil health deterioration, and stagnation or decline in crop yield over the past few decades (Bhatt et al. 2021).

A serious problem with this system of harvesting is crop residue burning which is rampant during the months of October–November (Gadde et al. 2009). North-West

N. Parwez · B. Biswas · S. Goel (✉)
Environmental Engineering and Management, Civil Engineering Department,
Indian Institute of Technology Kharagpur, Kharagpur 721 302, India
e-mail: sudhagiitkgp@gmail.com

S. Goel
Civil Engineering Department and School of Environmental Science and Engineering, Indian Institute of Technology Kharagpur, Kharagpur 721 302, India

S. Goel (ed.), *Advances in Solid and Hazardous Waste Management*,
https://doi.org/10.1007/978-3-031-49144-3_5

Fig. 1 Impact of paddy straw burning and smoke over North-West India on 18 November 2020. Available from https://earthobservatory.nasa.gov/images/147547/a-busy-season-for-crop-fires-in-northwestern-india

India is popularly known as the "breadbasket" of the country as two-thirds of the food grains are generated here (Chawala and Sandhu 2020). In this part of India, over 2.5 million farmers burn leftover rice straw in the field before preparing the fields for wheat seed plantation (Keil et al. 2020). A NASA Earth Observatory image obtained on 18 November 2020 using Suomi NPP—VIIRS and Terra—MODIS shows smoke over most of North-West India due to paddy straw burning (Fig. 1).

Open burning of crop residue in the fields increased from 18 million tons to 116 million tons from the year 1950–51 to 2017–18 (Government of India 2018; Economic Survey 2020). The Ministry of New and Renewable Energy estimated that 500 million tons of crop residue are generated annually in India (MoA 2014). In situ burning of crop residue has become extremely common due to a shortage of labour, cost of removing crop residue from the field and mechanized harvesting of crops (MoA 2014). Other reasons that make farmers choose to burn crop residue in situ are it is the cheapest method (Lopes et al. 2020), the residue has poor nutritional value (Singh and Sidhu 2014), the influence of other farmers, the increase in crop harvesting, time constraints between Kharif (monsoon) and Rabi (winter) season and lack of proper techniques and knowledge regarding residue management (Shyamsundar et al. 2019).

Crop residue burning has several detrimental effects which include loss of soil nutrients, damage to the soil ecosystem as the high temperature of burning destroys microbes and higher organisms, emission of gases like carbon dioxide, methane, carbon monoxide, nitrogen oxide, sulphur oxide and large amounts of particulate matter. Air pollution caused by paddy straw burning adversely affects the environment, and animal and human health. The US AQI value observed in Delhi on 9 November 2020 was 1039 which is extremely hazardous for health. Rice and wheat residue burning in India are responsible for more than 0.05% of the total greenhouse gas release into the environment (Gadde et al. 2009). The organic carbon pool and surface cover of the soil are affected by the burning of residues in the field, which leads to the reduction of food sources in the soil and an increase in soil erosion (Majumder and Kuzyakov 2010; Singh et al. 2020). Despite state government regulations and the "National Policy for Management of crop residues" implementation, the burning of crop residue in India continues.

Renewable energy can be generated from crop residues as it comprises hemicellulose, cellulose and lignin (Bhattacharyya and Barman 2018). As rice stubble is a low-density material and very bulky in nature, handling, transportation, and storage for energy generation becomes very difficult. The carbon budget of the ecosystem is significantly influenced by the crop residues produced by the RWCS system (Sharma et al. 2020; Vijayaprabhakar et al. 2017). Therefore, a transition from traditional crop residue management to an efficient, sustainable, and bio-economic technique with a minimum carbon footprint is required (Hosseini-Fashami et al. 2019; Venkatramanan et al. 2021). To address the above-mentioned problems associated with the existing practices involved in RWCS, this study aimed to characterize rice and wheat husk samples for major parameters like moisture content, nature of solids, calorific value, and CHNS analysis to understand the nature of the waste. Furthermore, some alternative techniques for the management of rice and wheat straw based on the results of the characterization are evaluated.

2 Material and Methods

2.1 Sample Collection

Wheat and rice are the two major sources of agricultural waste in India. Rice and wheat husk samples were collected from a rice mill in Jhargram on 19 Jan 2017. Both samples were collected in polyethylene bags and fastened properly to avoid any contamination during transportation and storage. The moisture content of the samples was determined immediately. Furthermore, the samples were dried and shredded properly in a mixer-grinder to obtain a homogenized sample for further analysis of important parameters like total solids, volatile solids, fixed solids, calorific value and elemental composition.

Table 1 Analytical methods and the parameters used for characterization of rice and wheat husk samples

Parameter	Method	References
Moisture content	Oven method	Worrell and Vesilind (2012)
Total solids (TS), Volatile solids (VS) and Fixed solids (FS)	Method 2540 G; gravimetric	APHA et al. (2012)
Calorific value	Bomb calorimeter	Worrell and Vesilind (2012)
Elemental (CHNS-O) analysis	Combustion method	Euro Vector EA3000 elemental analyzer

2.2 Sample Characterization

The samples were collected and evaluated for the following parameters, and the corresponding methods used for the characterization of waste are listed in Table 1. All samples were characterized in triplicate.

2.3 Power Potential of Biomass

Biomass quantities generated on an annual basis in the form of agricultural residues in the state of Punjab were estimated. Furthermore, the surplus biomass available for energy and power conversion was calculated. The methodology used in this study comprised various parameters such as the availability of crop residue, energy contents of the particular crop residues and the efficiency of power plants.

Theoretical Biomass Potential (TBP)

Theoretical Biomass Potential (TBP) represents the amount of straw produced within certain geographical boundaries per year in metric tons (MT) (Singh et al. 2008).

$$TBP = \sum_{i=1}^{n} R(i) \times P(i)$$

where, $P(i)$ is the yearly production of a specified crop within certain geographical boundaries in MT and $R(i)$ represents the residue-to-product ratio (RPR) of a particular crop. Total biomass potential is calculated by adding the values obtained for each crop. Representative values for the major crops are shown in Table 2.

Available Biomass Potential (ABP)

This represents the annual quantity of crop straw that is available for power generation in a particular region. To consider the constraints such as harvesting practices and other competing uses of the straw residue, an availability factor was introduced in the equation of theoretical biomass potential.

Table 2 Estimation of theoretical amount of crop residue in Punjab

Crop	Production (MT)	RPR[a]	Residue production (MT)[a]
Rice	10.25	1.5	15.38
Wheat	13.31	1.75	23.29
Maize	0.43	1.5	0.65
Cotton	0.37	3	1.1
Sugarcane	8.48	0.4	3.39

[a] Kumar et al. (2021)

$$ABP = \sum_{i=1}^{n} R(i) \times P(i) \times A(i)$$

where $A(i)$ is the accessibility or availability factor and its values are obtained from the literature. Representative values are shown in Table 3.

Available Energy Potential

This represents the amount of energy that can be harnessed using straw residue in a power generation plant. It can be calculated using the values of available potential and the calorific values of the straw residues.

$$AEP = \sum_{i=1}^{n} R(i) \times P(i) \times A(i) \times C(i)$$

where $C(i)$ is the calorific value of the individual crop residue that could be used in the power plant MJ/metric ton or MJ/T. AEP values were calculated for each crop and are shown in Table 4.

Available Energy and Power Potential of Biomass in Punjab

Operating hours for the plant are taken as 16 h per day. So, the total operating hours in a year are 5600 h considering that the plant runs 50 weeks.

Table 3 ABP values for different crops produced in Punjab

Crops	Residue production (MT)	Availability (%)	Available biomass potential—ABP (MT)
Rice	15.38	80	12.304
Wheat	23.29	30	6.987
Maize	0.65	80	0.52
Cotton	1.1	90	0.99
Sugarcane	3.39	40	1.356

Table 4 Estimation of available energy potential (AEP)

Crops	Available biomass potential (MT)	CV (MJ/T)[b]	AEP (MJ)
Rice	12.304	12784	1.57E+11
Wheat	6.987	12406	8.67E+10
Maize	0.52	14174	7.37E+09
Cotton	0.99	12946	1.28E+10
Sugarcane	1.356	12295	1.67E+10

CV: Calorific value
[b]Singh et al. (2021)

Table 5 Available power potential for different crops

Crops	Available power potential	
	Scenario 1: Eff = 20%	Scenario 2: Eff = 25%
Rice	1560	1951
Wheat	860	1075
Maize	73	91
Cotton	127	159
Sugarcane	165	207

Available Power Potential

The following equation can be used to calculate the available power generation potential of the power plant using straw residue as its feed. An efficiency factor and the operating time of the power plant are used as additional variables.

$$APP = \sum_{i=1}^{n} [R(i) \times P(i) \times A(i) \times C(i) \times \eta]/t$$

where η and t are the efficiency and the yearly operating hours of the plant, respectively. Generally, the efficiency of the power plant lies in the range of 20–40%. Operating hours are taken as 16 h per day and the plant is assumed to run for 50 weeks in the year. Calculated values for two scenarios: an efficiency of 20 and 25% are summarized in Table 5. Total power and exploitable power (in MW) for two different efficiencies—20 and 25% and for two different capacity factors—75 and 80% are summarized in Table 6.

2.4 Analytical Methods

The fundamental characteristics of any material are determined by proximate and ultimate analysis.

Table 6 Estimation of total and exploitable power (MW)

	Scenario 1: Eff = 20%, CF = 75%	Scenario 2: Eff = 20%, CF = 80%	Scenario 3: Eff = 25%, CF = 75%	Scenario 4: Eff = 25%, CF = 80%
Total power	2786	3637	3483	4546
Exploitable power	2090	2229	2612	2786

CF: Capacity factor; Eff: Efficiency

a. **Moisture content**: 25 g of each husk sample was taken in a crucible dish, and dried at 105 °C in an oven to a constant weight. A desiccator was used to cool the sample to room temperature, and the samples were weighed. The moisture content of each sample was determined by the percentage of weight lost in drying.
b. **Total Solids (TS), Fixed Solids (FS) and Volatile Solids (VS)**: Method 2540 G (APHA 2005) was used to determine the TS, FS and VS of the husk samples. Formulae used for the calculation of different solid contents are given below:

$$\text{Total Solids (in \%)} = [(a - b)/(c - b)] \times 100$$

$$\text{Fixed solids (in \%)} = [(d - b)/(a - b)] \times 100$$

$$\text{Volatile solids (in \%)} = [(a - d)/(a - b)] \times 100$$

where,

a Weight of dried residue and crucible;
b Weight of dried crucible;
c Weight of wet sample and crucible;
d Weight of residue and crucible after igniting.

c. **Calorific Value**: Calorific values of the samples were determined using a bomb calorimeter. The calibration of a bomb calorimeter has traditionally been called standardization. The calorimeter was calibrated by the combustion of benzoic acid, before the experiments.
d. **Elemental (C, H, N, S and O) analysis**: Euro Vector EA3000 Series of CHNS-O Elemental analyzer was used for the determination of the elemental composition of husk samples. The elemental composition was obtained in terms of percentage by dry weight. A tin capsule was used to weigh the samples, before dropping them into the combustible chamber of the analyzer. A GC column and a thermal conductivity detector (TCD) were used to separate and detect the elemental composition of the samples, respectively.

3 Results and Discussion

3.1 Sample Characterization

The wheat and rice husk samples were collected and analyzed for different chemical and physical parameters. The results are summarized and listed in Table 7.

All the samples were characterized in triplicates and the results are discussed here.

a. **Moisture content**: The moisture content of the samples was 9.58% for rice husk and 12.61% for wheat husk. The results of this study are in agreement with other studies which show the moisture content of rice and wheat samples ranged around 10% (Fermanelli et al. 2020; Raj et al. 2015).
b. **Total solids (TS)**: The total solid content for rice husk was found to be 90.42% while it was 87.39% for wheat samples. This corresponds to the higher moisture content of the wheat husk samples than the rice husk. This data is in accordance with earlier studies as well (Fermanelli et al. 2020; Rodríguez et al. 2021).
c. **Volatile Solids (VS)**: The volatile solids content for rice husk ranged from 91 to 95%, and for wheat husk, ranged from 96 to 99% with an average value of 93.03 and 97.56% respectively. Higher volatile solids are indicative of the presence of higher organic content in the samples. In general, because of their higher volatile content, biomass fuels are expected to ignite at lower temperatures than coal (Shen et al. 2017).
d. **Fixed Solids (FS)**: The fixed solid content for wheat ranges from 1 to 3%, while for rice, it ranges from 6 to 9%. The lower fixed solids of the samples correspond to the presence of lower inorganic content or inert material. Previous studies have shown that the volatile solid content is greater than the fixed solid content in wheat and rice samples as observed in other studies (Fermanelli et al. 2020; Raj et al. 2015).
e. **Calorific value**: The calorific values of rice and wheat husk samples were found to be 3938 cal/g and 4169 cal/g, respectively. The heat generated during combustion is represented by the calorific value.

Table 7 Characterization of rice and wheat husk samples

Parameters	Rice husk	Wheat husk
Moisture content (%)	9.58	12.61
Total solid (%)	90.42	87.39
Volatile solid (% of TS)	93.03	97.57
Fixed solid (% of TS)	6.97	2.43
Calorific value (Cal/g)	3938.28	4169.95
Total carbon (%C by wt.)	39.20	42.94
Total nitrogen (%N by wt.)	1.61	1.62
C/N ratio	28.44	30.84

f. **CHNSO elemental analysis**: The carbon content of rice and wheat husk was 39% for the rice husk and 43% for the wheat husk. Similarly, the hydrogen content was greater for wheat (8%) than for rice (6%). The higher value of hydrogen content in wheat husk indicates that it is likely to be more reactive than rice husk. The nitrogen contents of rice and wheat husk were not significantly different, i.e., 1.61 and 1.62%, respectively. For comparison, the nitrogen content of Indian coals varies within the range of 0.9–1.1%. During combustion, the nitrogen in the fuel will be converted to ammonia, elemental nitrogen and oxides of nitrogen depending on the conditions of the firing environment. The C/N ratios for both samples were moderately high (28.44 and 30.84, respectively). To achieve the final C/N ratio of 15–20 as required by composting standards, the samples will need supplements of nitrogen-rich material such as animal manure to make them suitable for composting.

3.2 Crop/Residue Generation and Power Generation Potential

Crop production data for the last five years were available from the Statistical Abstract of Punjab (2019). Wheat production was 13.31 million tons and rice production was 10.25 million tons. Figure 2 shows the percentages of various crops produced in Punjab. The highest production was for wheat at 40.49% followed by rice at 31.19%. Other major crops in descending order are sugarcane, cotton, maize, oilseeds like rapeseed, mustard, sunflower and groundnuts, followed by dals and bajra (millet).

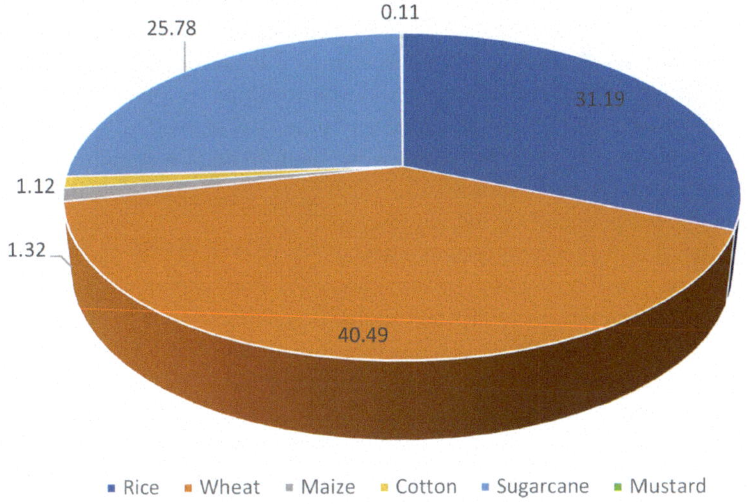

Fig. 2 Production of different crops in percentage

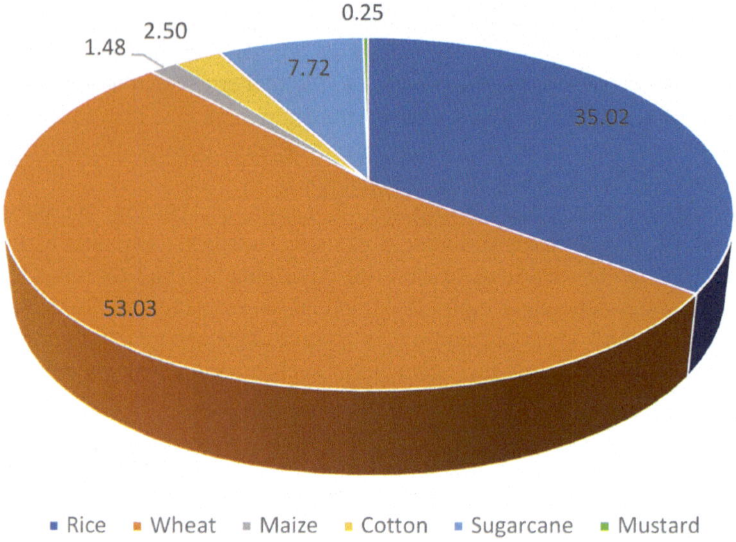

Fig. 3 Theoretical residue production of crops in percentage

Theoretical Biomass Potential: Incorporating the Residue-to-Product Ratio for the crops, the theoretical values of residue production from individual crops were calculated and are shown in Fig. 3. Based on the ratio values, the theoretical residue production is maximum for wheat followed by rice, while for other crops, the values (in million tons/year) are quite low and negligible.

Available Energy and Power Potential of Crops: Available energy and power potential of crop residue were calculated based on the availability of the crop residue and calorific values. The availability of rice straw was considered to be 80% while the availability of wheat residue was only 30% resulting in a major difference in the amount of available energy potential from the two crops. Different efficiency and capacity factors for the power plants were assumed and the results are shown in Figs. 4 and 5. The formulae shown in Sect. 2.3 were used to calculate the values of energy and power potential for individual crops.

The estimation of energy and power potential was done assuming that the available residue produced will be used for power generation. Two scenarios were considered for calculating the power potential of crops: the first with 20% power plant efficiency and the second with 25% efficiency of the power plant. The energy and power potential of rice residue was highest compared to other crops and was followed by wheat, sugarcane and cotton. Maize had the lowest power generation potential.

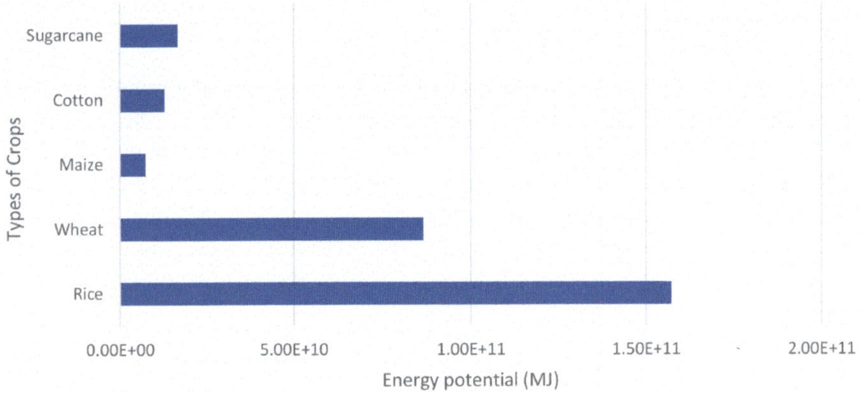

Fig. 4 Available energy potential of crops in Punjab in million Joules/year

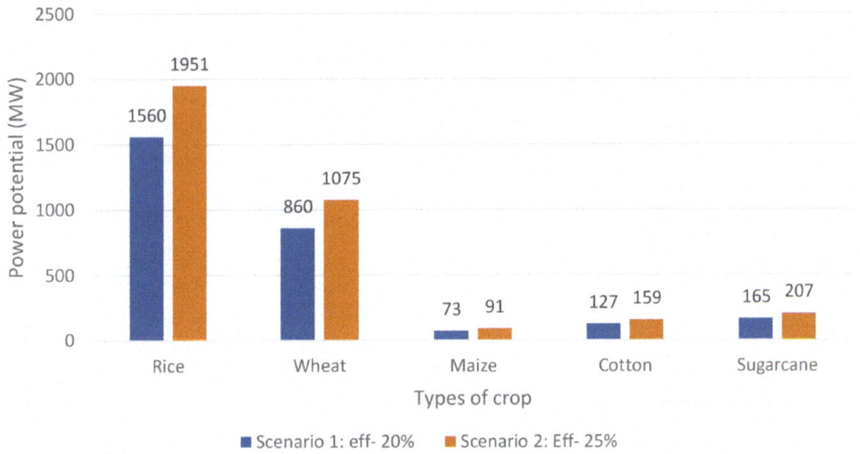

Fig. 5 Available power potential of crops in Punjab using two different efficiency values of 20 or 25%

3.3 Crop Residue Management

Crop residue can be reused in many different ways and this is a major challenge for researchers worldwide to come up with innovative and sustainable solutions for crop residue management. In situ management is the first set of options and includes soil mulching, baling/binding for use as a domestic or industrial fuel or to be used as fodder (MoA 2014). Ex situ options recommended by the Ministry of Agriculture include charcoal gasification, power generation by incineration or biofuel (biogas or bioethanol) production, packaging material, used in paper, board and panel manufacture, composting and mushroom cultivation.

3.3.1 Composting

Composting is the process of conversion of organic material under humid, hot, and aerobic conditions. In this process, the volume of organic waste can be reduced to 50–60% of the original volume. The end product after the microbial conversion is called compost which can serve as a soil conditioner and as a fertilizer supplement. The waste is decomposed by microorganisms like bacteria and fungi as well as higher organisms like earthworms, beetles and other insects, etc. Microorganisms play a significant role in composting as they decompose and mineralize agricultural wastes.

The efficiency of the composting process and the quality of compost generated depends on several factors such as particle size, ambient temperature, initial pH, nutrient (macronutrients and micronutrients) availability, trace metals like Ca, Cu, Co, Mg, Mn and others, C/N ratio, moisture content (50–60%, optimum—55%), organic content, seed (bacteria, fungi, earthworms, beetles, etc.), air or O_2 availability which is determined by the degree of mixing or turning, and land requirement for the composting process.

The ideal C:N ratio for composting should be 25–35 (Masters and Ela 2008). High C:N ratios inhibit the growth of microbes while low C:N ratios enhance biodegradation rates. Generally, crop residue has high C:N ratios, i.e., low N concentration, and it is necessary to combine these materials with the material having a low C:N ratio to create an ideal mix.

The C/N ratio of the crop residue in the samples analyzed in this study was in the range of 25–30, which is close to an ideal value for effective composting. The rice and wheat stubble can be used in cattle sheds as bed material and further processed for composting. About 2–3 kg of animal urine can be absorbed in each kg of paddy straw, and it can be composted on the farm (in situ composting). The strength-weakness-opportunity-threat (SWOT) of the composting process is shown in Fig. 6.

Several methods of composting can be used for crop residue management and these include the Indore Method, Nadep Method, Coimbatore and Bangalore Methods. These are indigenous methods developed for Indian conditions and are suitable for combining animal manure which is rich in N with crop residue which has low N content. A detailed description of the Indore Method is provided in Chap. 4 and detailed descriptions of other methods can be found on the Internet.

3.3.2 Incineration

The complete combustion of wastes in the presence of air and in a controlled system is termed incineration. It leads to the complete destruction of waste as the temperature of the incinerator is very high and varies between 980 and 2000 °C. Incineration is the most common thermal treatment as it can reduce the volume of solid waste to 10–20% of its original volume. In the process of incineration, the waste is converted to CO_2, water vapour, and ash. This system can be used as a way of recycling energy that

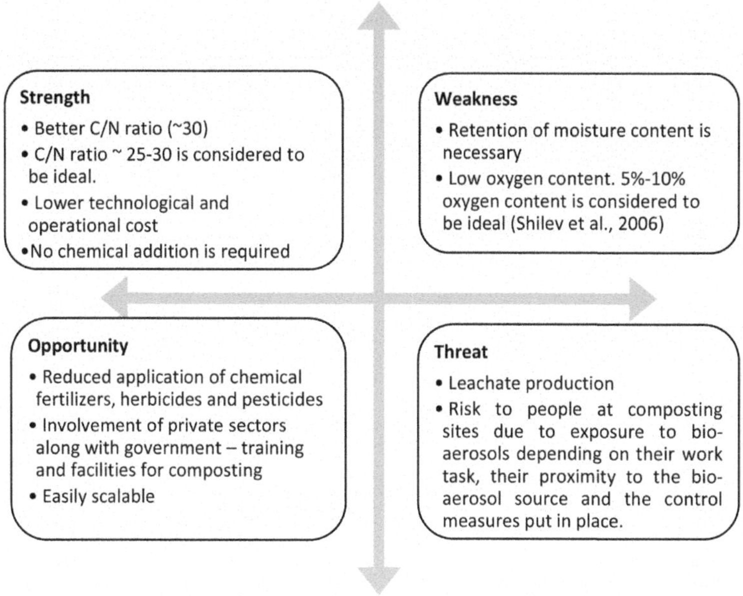

Fig. 6 SWOT analysis of composting

can be used for generating power. A strength-weakness-opportunity-threat (SWOT) analysis of the incineration process was assessed and shown in Fig. 7.

3.3.3 Biofuel Generation

Biopower can be produced from crop residue, through bacterial action, liquid fuel, and burning. Power plants based on crop residues are preferred by farmers (Chauhan et al. 2012). Energy production from biomass shows significant potential for the management of crop residues due to:

(a) Large amounts of crop residue generated per annum and are expected to increase with the increasing demand for food,
(b) Bioenergy production facilities can be very useful in employment generation and are a step closer to a circular economy, and
(c) The spatio-temporal differences in the generation of crop residue can be used for creating decentralized energy production facilities.

3.3.4 Residue Retention Techniques

Crop residue can be retained on the field and needs to be mulched into the soil. There are numerous benefits of crop retention on the field surface such as:

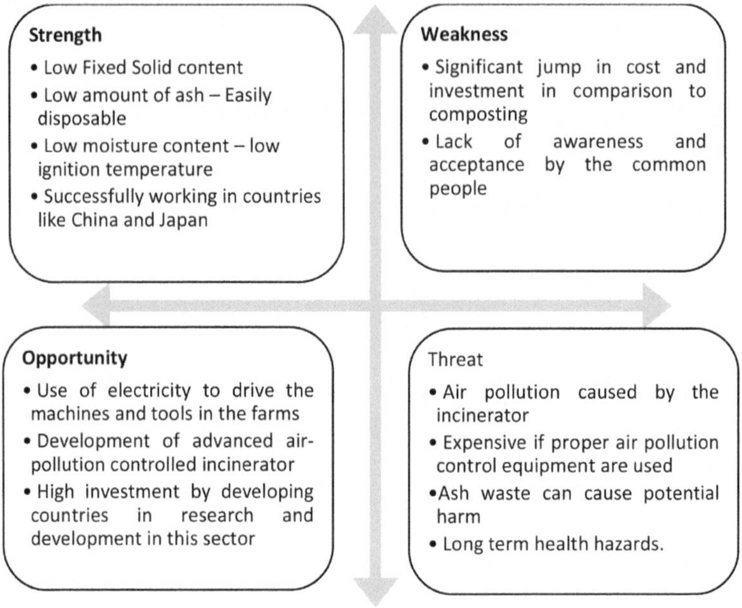

Fig. 7 SWOT analysis of incineration

(a) Crop residues release bases like hydroxyls which play a significant role in soil health by maintaining soil pH and preventing the soil from becoming acidic,
(b) It reduces the application of herbicides as less weed growth has been observed,
(c) Plant nutrients are recycled, thus helping in carbon sequestration (Jain et al. 2014),
(d) Remains of residue carbon result in immobilization of applied nitrogen (Chauhan et al. 2012),
(e) Reduces use of fertilizers, as the residue act as a nutrient pool for plants, and
(f) They help to maintain and improve the soil ecosystem.

Soil mulching can be done either manually or mechanically. Due to labour shortages, manual mulching is difficult in Punjab and farmers and the government are using machines for mulching. Another method that is in use in the Delhi region is the Pusa Decomposer (PD) where a mixture of nutrients and bio-inoculum is sprayed on the crop residue in situ. The residue is decomposed by the bio-inoculum and nutrients released from the residue are added to the soil, thus, enriching it.

Farm machinery developed based on resource conservation promises a better future in the management of crop residues, as it helps in achieving sustainable agriculture, improving crop productivity, soil health, and reducing pollution (Jat et al. 2009; Palm et al. 2014). These technologies are extremely valuable in crop residue management.

The Happy Seeder (HS) machine consists of a unit for straw management and a unit for sowing (Lohan et al. 2018). It is a tractor-powered machine—cuts and lifts

Fig. 8 Turbo happy seeder (Ref: By Father of God—Own work, CC BY-SA 4.0, https://commons.wikimedia.org/w/index.php?curid=101089993)

rice residue, sows into the bare soil and deposits the stubble over the sown area as mulch, and can cover up to 0.3–0.4 ha. This machine can combine the functions of stubble mulching, seed, and fertilizer drilling operations in a single pass. The HS technology is found to have significant positive effects on the environment and community as well as on individual farmers. It can be a viable alternative to burning as a means of managing rice residues before sowing wheat in farms in Punjab. Field trials have indicated that it improves crop yields, decreases weed growth, improves soil quality, and reduces water consumption. Figure 8 shows a turbo HS working on the field in Punjab after rice harvesting.

The use of HS technology has increased from 8 to 600 ha of land area from the year 2006–07 to 2014–15 in India (Lohan et al. 2018). The HS provides an economic and practical solution to the rice residue burning problem in India. Many studies indicate that crop yield has been maintained or improved using this technology. The HS machines are provided to all farmers at an 80% subsidy to farmer groups and 50% to individual farmers. Despite this, some farmers are reluctant to adopt this technology for two reasons: the machines are not available to all and there is an initial decline in wheat yield until management is improved. Paddy straw has to be spread evenly on the field and thick bunches of straw make use of HS difficult.[1] HS allows wheat to be sown immediately after the rice has been harvested. This helps to avoid pre-sowing irrigation and the paddy or rice straw from the previous harvest is mulched into the soil using HS. A SWOT analysis of the residue retention techniques was conducted and is shown in Fig. 9. This information will help to identify the total benefits of this process for successful adoption.

[1] https://indianexpress.com/article/explained/explained-using-happy-seeder-and-how-it-affects-wheat-yield-6017640/.

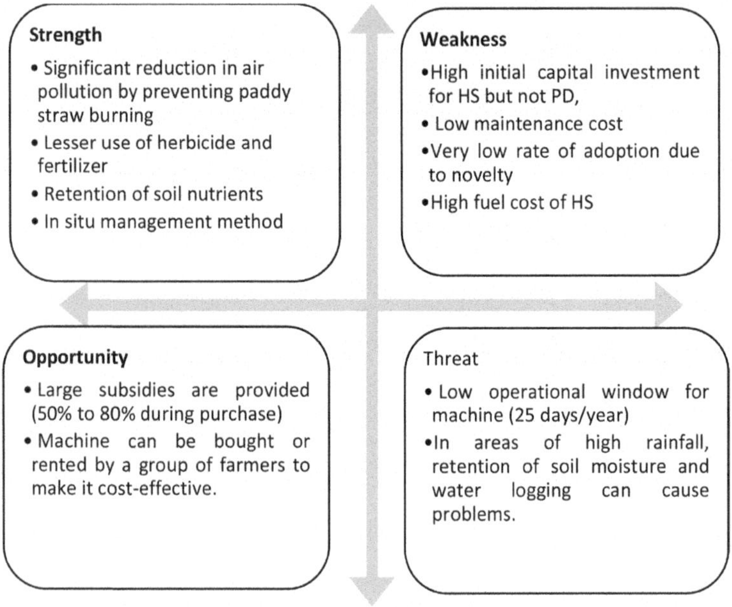

Fig. 9 SWOT analysis of residue retention

4 Conclusions

In conclusion, crop residue is the biomass that is abandoned in the field after harvesting. These residues have great value for the generation of fuel, livestock feed, raw material for composting and for maintaining the nutrient pool in the field. Crop residue management remains an unresolved issue because of the ignorance of farmers about the adverse effects of crop residue burning on the environment, human health, and soil health. The main harvested crops in Punjab are wheat and rice, and crop residue management was the main focus of the current research. A few residue management techniques were discussed in detail, and their strengths and weaknesses were evaluated. Crop residue management has to shift from the traditional practice of crop residue burning to more sustainable options keeping the circular economy in mind. Sustainable practices provide great opportunities to resolve the problem of open-field burning of crop residue. This study found great potential for the implementation of different sustainable options like composting, biofuel generation, microbial decomposition or Happy Seeder technology in the study region of Punjab. Implementation of these processes will reduce the environmental damage caused by open-field burning of crop residue and will achieve many benefits.

References

APHA, AWWA & WEF (2012). *Standard Methods for the Examination of Water and Wastewater* (22nd ed.). APHA.

Bhatt, R., Singh, P., Hossain, A. and J. Timsina (2021). Rice–wheat system in the northwest Indo-Gangetic plains of South Asia: Issues and technological interventions for increasing productivity and sustainability. *Paddy and Water Environment*, **19**(**3**): 345–365. https://doi.org/10.1007/s10333-021-00846-7.

Bhattacharyya, P. and D. Barman (2018). Crop Residue Management and Greenhouse Gases Emissions in Tropical Rice Lands. *In:* María Ángeles Muñoz, M.A. and R. Zornoza (eds.). Soil Management and Climate Change (pp. 323–335). Elsevier. https://doi.org/10.1016/B978-0-12-812128-3.00021-5.

Chauhan, B.S., Mahajan, G., Sardana, V., Timsina, J. and M.L. Jat (2012). Productivity and sustainability of the rice–wheat cropping system in the Indo-Gangetic plains of the Indian subcontinent: Problems, opportunities, and strategies. *Advances in Agronomy*, **117**: 315–369. https://doi.org/10.1016/B978-0-12-394278-4.00006-4.

Chawala, P. and H.A.S. Sandhu (2020). Stubble burn area estimation and its impact on ambient air quality of Patiala & Ludhiana district, Punjab, India. *Heliyon*, **6**(**1**): e03095. https://doi.org/10.1016/j.heliyon.2019.e03095.

Fermanelli, C.S., Córdoba, A., Pierella, L.B. and C. Saux (2020). Pyrolysis and copyrolysis of three lignocellulosic biomass residues from the agro-food industry: A comparative study. *Waste Management*, **102**: 362–370. https://doi.org/10.1016/j.wasman.2019.10.057.

Gadde, B., Bonnet, S., Menke, C., & Garivait, S. (2009). Air pollutant emissions from rice straw open field burning in India, Thailand and the Philippines. *Environmental Pollution, 157*(5), 1554–1558. https://doi.org/10.1016/j.envpol.2009.01.004.

Gathala, M.K., Kumar, V., Sharma, P.C., Saharawat, Y.S., Jat, H. S., Singh, M., Kumar, A., Jat, M.L., Humphreys, E., Sharma, D.K., Sharma, S. and J.K. Ladha (2013). Optimizing intensive cereal-based cropping systems addressing current and future drivers of agricultural change in the northwestern Indo-Gangetic Plains of India. *Agriculture, Ecosystems and Environment*, **177**: 85–97. https://doi.org/10.1016/j.agee.2013.06.002.

Government of India (2018). Agricultural Statistics at a Glance 2018. Ministry of Agriculture and Farmers' Welfare, New Delhi. Downloaded from https://agricoop.gov.in.

Government of India (2020). Economic Survey 2020–21. Statistical Appendix, Downloaded from https://indiabudget.gov.in.

Hosseini-Fashami, F., Motevali, A., Nabavi-Pelesaraei, A., Hashemi, S.J. and C. Kwok-wing (2019). Energy-Life cycle assessment on applying solar technologies for greenhouse strawberry production. *Renewable and Sustainable Energy Reviews*, **116**: 109411. https://doi.org/10.1016/j.rser.2019.109411.

Jain, N., Pathak, H. and A. Bhatia (2014). Sustainable management of crop residues in India. *Current Advances in Agricultural Sciences*, **6**(**1**): 1–9.

Jat, M.L., Gathala, M.K., Ladha, J.K., Saharawat, Y.S., Jat, A.S., Kumar, V., Sharma, S. K., Kumar, V. and R. Gupta (2009). Evaluation of precision land leveling and double zero-till systems in the rice–wheat rotation: Water use, productivity, profitability and soil physical properties. *Soil and Tillage Research*, **105**(**1**): 112–121. https://doi.org/10.1016/J.STILL.2009.06.003.

Keil, A., Krishnapriya, P.P., Mitra, A., Jat, M.L., Sidhu, H.S., Krishna, V.V. and P. Shyamsundar (2020). Changing agricultural stubble burning practices in the Indo-Gangetic plains: Is the Happy Seeder a profitable alternative? *International Journal of Agricultural Sustainability*, **19**(**2**): 128–151. https://doi.org/10.1080/14735903.2020.1834277.

Kumar, S., Mangaraj, P., Beig, G., Samal, A., Pradhan, C., Dash, S. and B. Tyagi (2021). Quantifying the high resolution seasonal emission of air pollutants from crop residue burning in India * Intergovernmental Panel on Climate Change. *Environmental Pollution*, **286**: 117165. https://doi.org/10.1016/j.envpol.2021.117165.

Lohan, S.K., Jat, H.S., Yadav, A.K., Sidhu, H.S., Jat, M.L., Choudhary, M., Peter, J.K. and P.C. Sharma (2018). Burning issues of paddy residue management in north-west states of India. *Renewable and Sustainable Energy Reviews*, **81**: 693–706. https://doi.org/10.1016/J.RSER.2017.08.057.

Lopes, A.A., Viriyavipart, A. and D. Tasneem (2020). The role of social influence in crop residue management: Evidence from Northern India. *Ecological Economics*, **169**: 106563. https://doi.org/10.1016/J.ECOLECON.2019.106563.

Majumder, B. and Y. Kuzyakov (2010). Effect of fertilization on decomposition of 14C labelled plant residues and their incorporation into soil aggregates. *Soil and Tillage Research*, **109(2)**: 94–102. https://doi.org/10.1016/j.still.2010.05.003.

Masters, G. and W. Ela (2008). *Introduction to Environmental Engineering and Science*. Prentice Hall.

Palm, C., Blanco-Canqui, H., DeClerck, F., Gatere, L. and P. Grace (2014). Conservation agriculture and ecosystem services: An overview. *Agriculture, Ecosystems & Environment*, **187**: 87–105. https://doi.org/10.1016/J.AGEE.2013.10.010.

Pathak, H., Singh, R., Bhatia, A. and N. Jain (2006). Recycling of rice straw to improve wheat yield and soil fertility and reduce atmospheric pollution. *Paddy and Water Environment*, **4(2)**: 111–117. https://doi.org/10.1007/s10333-006-0038-6.

Raj, T., Kapoor, M., Gaur, R., Christopher, J., Lamba, B., Tuli, D.K. and R. Kumar (2015). Physical and chemical characterization of various indian agriculture residues for biofuels production. *Energy and Fuels*, **29(5)**: 3111–3118. https://doi.org/10.1021/ef5027373.

Rodríguez, A., Gea, T., Sánchez, A. and X. Font (2021). Agro-wastes and inert materials as supports for the production of biosurfactants by solid-state fermentation. *Waste and Biomass Valorization*, **12(4)**: 1963–1976. https://doi.org/10.1007/s12649-020-01148-5.

Saharawat, Y.S., Singh, B., Malik, R.K., Ladha, J.K., Gathala, M., Jat, M.L. and V. Kumar (2010). Evaluation of alternative tillage and crop establishment methods in a rice-wheat rotation in North Western IGP. *Field Crops Research*, **116(3)**: 260–267. https://doi.org/10.1016/j.fcr.2010.01.003.

Sharma, S., Singh, P. and S. Kumar (2020). Responses of soil carbon pools, enzymatic activity, and crop yields to nitrogen and straw incorporation in a rice-wheat cropping system in North-Western India. *Frontiers in Sustainable Food Systems*, **4**: 203. https://doi.org/10.3389/fsufs.2020.532704.

Shen, Z., Zhang, Y., Jin, F., McMillan, O. and A. Al-Tabbaa (2017). Qualitative and quantitative characterisation of adsorption mechanisms of lead on four biochars. *Science of the Total Environment*, **609**: 1401–1410. https://doi.org/10.1016/j.scitotenv.2017.08.008.

Shyamsundar, P., Springer, N.P., Tallis, H., Polasky, S., Jat, M.L., Sidhu, H.S., Krishnapriya, P.P., Skiba, N., Ginn, W., Ahuja, V., et al. (2019). Fields on fire: Alternatives to crop residue burning in India. *Science*, **365(6453)**: 536–538. https://doi.org/10.1126/science.aaw4085.

Singh, J., Panesar, B.S. and S.K. Sharma (2008). Energy potential through agricultural biomass using geographical information system-A case study of Punjab. *Biomass and Bioenergy*, **32(4)**: 301–307. https://doi.org/10.1016/j.biombioe.2007.10.003.

Singh, Y. and H.S. Sidhu (2014). Management of cereal crop residues for sustainable rice-wheat production system in the Indo-Gangetic Plains of India. *Proceedings of the Indian National Science Academy*, **80(1)**: 95–114. https://doi.org/10.16943/ptinsa/2014/v80i1/55089.

Singh, P., Singh, G. and G.P.S. Sodhi (2020). Energy and carbon footprints of wheat establishment following different rice residue management strategies vis-à-vis conventional tillage coupled with rice residue burning in north-western India. *Energy*, **200**: 117554. https://doi.org/10.1016/j.energy.2020.117554.

Singh, M., Singh Brar, Y. and H. Singh (2021). Critical assessment of biomass material for power generation in Punjab, India. *Materials Today: Proceedings*, **48(5)**: 927–931. https://doi.org/10.1016/j.matpr.2021.04.414.

Venkatramanan, V., Shah, S., Rai, A.K. and R. Prasad (2021). Nexus between crop residue burning, bioeconomy and sustainable development goals over North-Western India. *Frontiers in Energy Research*, **8**: 392. https://doi.org/10.3389/fenrg.2020.614212.

Vijayaprabhakar, A., Nalliah Durairaj, S. and J. Vinoth Raj (2017). Effect of rice straw management options on soil available macro and micro nutrients in succeeding rice field. *International Journal of Chemical Studies*, **5**(**4**): 410–413.

Wikipedia (2021). Land use statistics by country, https://en.wikipedia.org/wiki/Land_use_statistics_by_country.

Worrell, W.A. and P.A. Vesilind (2012). Solid Waste Engineering. Cengage Learning, USA.

Chapter 6
Polycyclic Aromatic Hydrocarbons (PAH) Contamination of Soil in Informal E-Waste Recycling Facilities

Deblina Dutta and Sudha Goel

1 Introduction

Technological innovations and the rapid growth of the electronics industry have led to the proliferation of electronic waste (e-waste), including obsolete refrigerators, washing machines, mobile phones, computers, printers, televisions, and other appliances. E-waste recycling is useful for reclaiming valuable metals, including gold, silver, palladium, and copper, and is more efficient in yielding such metals than conventional metal mines (Goosey and Kellner 2003; Xiang et al. 2007; Sun et al. 2016). However, besides toxic metals such as chromium, arsenic, cadmium, lead, and mercury, organic pollutants such as polycyclic aromatic hydrocarbons (PAHs), polychlorinated biphenyls, and polybrominated diphenyl ethers (PBDEs) are released into the environment during e-waste recycling and disposal due to the crude methods used, thus, causing serious risks to local ecosystems and human health (Gioia et al. 2011; Sankhla et al. 2016).

PAHs are organic contaminants that are composed of fused aromatic rings of carbon and hydrogen atoms; these are derived mainly from the incomplete combustion of organic material (Wang et al. 2012). There are 16 PAHs that are listed as priority pollutants by the USEPA and are shown in Table 1 along with their chemical formulae, molecular weights, USEPA rating, reference dose (RfDo) and cancer slope factors (CSFo) for the oral (ingestion) route of exposure (Mojiri et al. 2019; Manoli and Samara 1999; USEPA 2004). All 16 PAHs are relatively insoluble in water and have K_{ow} values ranging from 106 to 7×10^6. High K_{ow} values are indicative of the

D. Dutta (✉)
Department of Environmental Science and Engineering, School of Engineering and Sciences, SRM University-AP, Amaravati, Andhra Pradesh 522 240, India
e-mail: deblina69envs@gmail.com

S. Goel
Civil Engineering Department and School of Environmental Science and Engineering, Indian Institute of Technology Kharagpur, Kharagpur 721 302, India

S. Goel (ed.), *Advances in Solid and Hazardous Waste Management*,
https://doi.org/10.1007/978-3-031-49144-3_6

bioaccumulation potential of the compound and its subsequent toxicity to humans and ecosystems (Sawyer et al. 2003).

PAHs are ubiquitous in nature and can persist for a long time in the environment. They are known to be genotoxic and carcinogenic. In recent years, air, soil, and water surrounding e-waste dismantling areas were found to be heavily contaminated by heavy metals and PAHs (Daso et al. 2016; Luo et al. 2015).

Nearly 50–80% of all e-waste generated in developed countries is exported to developing countries such as China, India, Pakistan, Vietnam, and West African countries due to low labour costs and poor implementation of environmental regulations (Asante et al. 2011; Sthiannopkao and Wong 2013). To date, approximately 70% of the e-waste generated worldwide is processed in China (Robinson 2009), amounting to approximately 28 million tons per year (Zhang et al. 2012). E-waste is often incinerated by workers in the informal sector at low temperatures which results in the generation of polycyclic aromatic hydrocarbons (PAHs) (Adaramodu et al. 2012; Wei and Liu 2012; Yu et al. 2006). Soil samples in and around the dumping sites are frequently found to be contaminated with PAHs (Leung et al. 2015; Shi et al. 2016). Soil is an important environmental medium that can absorb and accumulate pollutants and cause contamination after the major pollution sources are controlled (Wong et al. 2007). Ingestion of these PAHs by humans and other organisms can result in toxic effects. Prolonged and continuous exposure affects human beings and gives

Table 1 Classification of PAHs

Sl. No	PAHs	Formula	Molecular weight (g/mol)	USEPA rating	Rfdo (mg/kg/day)	CSFo (mg/kg/day)$^{-1}$
1	Acenaphthene	C12H10	154.2	No class	6.00E-02	
2	Acenaphthylene	C12H8	152.1	D		
3	Anthracene	C14H10	178.2	D	3.00E-01	
4	Benzo[a]anthracene	C18H12	228.3	B2		1.00E-01
5	Benzo[b]fluoranthene	C20H12	252.3	B2		1.00E-01
6	Benzo[k]fluoranthene	C20H12	252.3	B2		1.00E-02
7	Benzo[a]pyrene	C20H12	252.3	B2	3.00E-04	1.00E+00
8	Benzo[ghi]perylene	C22H12	276.3	D	2.00E-03	
9	Chrysene	C18H12	228.2	B2		1.00E-03
10	Dibenzo[a,h]anthracene	C22H14	278.3	B2		1.00E+00
11	Fluoranthene	C16H10	202.2	D	4.00E-02	
12	Fluorene	C13H10	166.2	D	4.00E-02	
13	Indeno[1,2,3]pyrene	C22H12	276.3	B2		1.00E-01
14	Naphthalene	C10H8	128.1	C	1.00E-01	
15	Phenanthrene	C14H10	174.2	D	3.00E-02	
16	Pyrene	C16H10	202.2	D	3.00E-02	

Table 2 Pollution classes for surface soils based on total PAHs (mg/kg)

Total PAH content	Pollution class	Soil assessment
<0.2	0	Not polluted
0.2–0.6	I	Very slightly polluted
0.6–1.0	II	Slightly polluted
1.0–5.0	III	Polluted
5.0–10.0	IV	Highly polluted
>10.0	V	Very highly polluted

rise to carcinogenic and non-carcinogenic effects (Kamunda et al. 2016). The presence of PAHs in the soil is well-documented in the literature and health risks due to exposure to PAHs have also been assessed and reported (Deng et al. 2006; Goel 2006; Wang et al. 2012; Luo et al. 2015; Gu et al. 2019; Chen et al. 2019; Liu et al. 2020).

2 Pollution Class and Standards of PAHs

Pollution classes for total PAHs in surface soil are provided in Table 2 (Kumar et al. 2017). ATSDR standards are noted in Table 3.[1] USEPA has standards for some PAHs in the workplace and for water and these are shown in Table 3.

3 Effects of PAHs Released from Incineration of E-Waste

PAHs are known to be lipophilic due to which they can enter the food chain and also get absorbed into the skin (Wang et al. 2012). PAHs are of two types: low molecular weight with 2–3 benzene rings and high molecular weight having 4–7 benzene rings. The low molecular weight PAHs (naphthalene, phenanthrene, fluorene, and anthracene) are acutely toxic to aquatic life in comparison to high molecular weight PAHs (chrysene and coronene). High molecular weight PAHs are not acutely toxic but carcinogenic (Adeyi and Oyeleke 2017; Nasher et al. 2013). PAHs are hard to degrade abiotically and biotically resulting in their persistence in the environment (Yang et al. 2019).

Shen et al. (2013) studied the Global Atmospheric Emissions of Polycyclic Aromatic Hydrocarbons from 1960 to 2008 and reported that globally the atmospheric emission of 16 PAHs was 504 Gg in 2007 annually. The biggest contributor of PAHs in the world is Asia (South, East and South-east) (Miura et al. 2019; Shen et al. 2013). While the bulk of PAHs is emitted during biomass burning, e-waste incineration in the open can release significant amounts of PAHs in the air and soil. The informal sector in India generally burns e-waste in the open for separating different

[1] https://www.atsdr.cdc.gov/csem/csem.asp?csem=13&po=8.

Table 3 Standards for PAHs

Agency	Focus	Level	Comments
American conference of governmental industrial hygienists	Air: workplace	0.2 mg/m^3 for benzene-soluble coal tar pitch fraction	8-h workday
National institute for occupational safety and health	Air: workplace	0.1 mg/m^3 for coal tar pitch volatile agents A	Exposure limit of 10 h
Occupational safety and health administration	Air: workplace	0.2 mg/m^3 for benzene soluble coal tar pitch fraction	Permissible exposure limit of 8 h
U.S. Environmental Protection Agency (USEPA)	Water	0.0001 mg/L	Maximum contaminant level for benzo(a) anthracene
USEPA	Water	0.0002 mg/L	Maximum contaminant level for benzo(a) pyrene, benzo(b)fluoranthene, benzo(k) uoranthene, chrysene
USEPA	Water	0.0003 mg/L	Maximum contaminant level for dibenzo (a, h) anthracene
USEPA	Water	0.0004 mg/L	Maximum contaminant level for indenol(1,2,3-c,d) pyrene

components of e-waste. Several case studies which showed the presence of PAHs in and around e-waste dumping sites are summarized in Table 4.

4 Conclusions

The present e-waste recycling processes used in developing countries are extremely hazardous and have adverse impacts on human health and the environment. PAHs are released mainly during the open burning of e-waste thereby contaminating soil, air and water. As these pollutants spread through soil, air and water, they can bioconcentrate and bioaccumulate in the food chain. To mitigate the challenges related to informal e-waste recycling facilities, adequate and systematic international effort is required to reduce pollution and health hazards associated with cheap and crude e-waste recycling methods.

Table 4 Case studies related to PAHs detected in soil samples from e-waste dumping sites

Study location	PAHs found	Range of concentration	References
Beijing, Tianjin, North China	All 16 PAHs are present	31.6–1475 ng/g	Wang et al. (2010)
Zeguo, Taizhou area, China	Phenanthrene, anthracene, fluoranthene and pyrene	Total PAH-262.6 to 3,420.2 μg/kg	Tang et al. (2010)
Longtang Town, Guangdong Province, South China	Phenanthrene, fluoranthene, pyrene, and benzo[a]anthracene	Soil-127 to 10600 ng/g Plant-199 to 2420 ng/g	Wang et al. (2012)
Guangzhou in southern China	Acenaphthene, fluorene, phenanthrene, anthracene, fluoranthene, pyrene, benzo[a]anthracene, chrysene, benzo[b]fluoranthene, benzo[k]	Atmosphere at all heights-15.1 to 17.7 and 5.8–6.7 ng/m^3	Luo et al. (2015)
	Fluoranthene, benzo[e]pyrene, benzo[a]pyrene, indeno[1,2,3-cd]pyrene, dibenzo[a,h]anthrancene, benzo[g,h,i]perylene, coronene		
Guiyu, China	All 16 PAHs are present	Soil-56 to 567 ng/g Sediment-181 to 3034 ng/g	Xu et al. (2016)
Gwangju City, Korea	Naphthalene, 2-methyl naphthalene, 1-methyl naphthalene, Phenanthrene, Anthracene, Fluoranthene, Pyrene, Chrysene, Benzo[k]fluoranthene, Benzo[a]pyrene, Dibenz[a, h]anthracene, Benzo[g,h,i]pyrene	120–273 ng/g	Islam et al. (2017)
Vietnam, Philippines and Ghana	15 PAHs, 26 Cl-PAHs and 14 Br-PAHs	PAH-390 and 710 ng/g (Ghana); 2900 and 4000 ng/g (Philippines); 7200 and 6500 ng/g (Vietnam) Cl-PAH-21 to 2800 ng/g Br-PAH-5.8 to 520 ng/g	Nishimura et al. (2017)
Lagos and Ibadan, Nigeria	Total PAHs	1,756–2,224 μg/kg at 0–15 cm level, 1,664–2,152 μg/kg at 15–30 cm	Adeyi and Oyeleke (2017)

(continued)

Table 4 (continued)

Study location	PAHs found	Range of concentration	References
Xinqiao, China	Naphthalene, acenaphthylene, acenaphthene, fluorine, phenanthrene, anthracene, fluoranthene, pyrene, chrysene, benzo[a]anthracene, benzo[b]	LPAHs—85.41 μg/kg HPAHs—328.64 μg/kg	Gu et al. (2019)
	Fluoranthene, benzo[k]fluoranthene, benzo[a]pyrene; dibenzo[a,h]anthracene, benzo[g,h,I]perylene, indeno[1,2,3-cd]pyrene		
South China	PAHs, chlorinated PAHs (ClPAHs), brominated PAHs (BrPAHs), oxy-PAHs (OPAHs)	PAHs-1.53 $\times$ 10^4–2.02 $\times$ 10^5, ClPAHs-32.3–364, BrPAHs-8.29–1.13 $\times$ 10^3, OPAHs–923–1.39 $\times$ 10^4	Liu et al. (2020)
Accra (Ghana)	All PAHs	16 USEPA PAHs-850 to 10,000 ng/g All PAHs-1,300 to 16,000 ng/g	Moeckel et al. (2020)

References

Adaramodu, A.A., Osuntogun, A.O. and C.O. Ehi-Eromosele (2012). Heavy metal concentration of surface dust present in e-waste components: The Westminister Electronic Market, Lagos case study. *Resources and Environment*, **2(2)**: 9–13. https://doi.org/10.5923/j.re.20120202.02.

Adeyi, A.A. and P. Oyeleke (2017). Heavy metals and polycyclic aromatic hydrocarbons in soil from e-waste dumpsites in Lagos and Ibadan, Nigeria. *Journal of Health and Pollution*, **7(15)**: 71–84. https://doi.org/10.5696/2156-9614-7.15.71.

Asante, K.A., Adu-Kumi, S., Nakahiro, K., Takahashi, S., Isobe, T., Sudaryanto, A., Devanathan, G., Clarke, E., Ansa-Asare, O.D., Dapaah-Siakwan, S. and S. Tanabe (2011). Human exposure to PCBs, PBDEs and HBCDs in Ghana: Temporal variation, sources of exposure and estimation of daily intakes by infants. *Environment International*, **37(5)**: 921–928. https://doi.org/10.1016/j.envint.2011.03.011.

Chen, H., Ma, S., Yu, Y., Liu, R., Li, G., Huang, H. and T. An (2019). Seasonal profiles of atmospheric PAHs in an e-waste dismantling area and their associated health risk considering bioaccessible PAHs in the human lung. *Science of the Total Environment*, **683**: 371–379. https://doi.org/10.1016/j.scitotenv.2019.04.385.

Daso, A.P., Akortia, E. and J.O. Okonkwo (2016). Concentration profiles, source apportionment and risk assessment of polycyclic aromatic hydrocarbons (PAHs) in dumpsite soils from Agbogbloshie e-waste dismantling site, Accra, Ghana. *Environmental Science and Pollution Research*, **23(11)**: 10883–10894. https://doi.org/10.1007/s11356-016-6311-3.

Deng, W.J., Louie, P.K.K., Liu, W.K., Bi, X.H., Fu, J.M. and M.H. Wong (2006). Atmospheric levels and cytotoxicity of PAHs and heavy metals in TSP and PM2.5 at an electronic waste recycling site in southeast China. *Atmospheric Environment*, **40(36)**: 6945–6955. https://doi.org/10.1016/j.atmosenv.2006.06.032.

Gioia, R., Eckhardt, S., Breivik, K., Jaward, F.M., Prieto, A., Nizzetto, L. and K.C. Jones (2011). Evidence for major emissions of PCBs in the West African region. *Environmental Science & Technology*, **45**(**4**): 1349–1355. https://doi.org/10.1021/es1025239.

Goel, S. (2006). Health risk assessment for a contaminated site: A case study. P*ractice Periodical of Hazardous, Toxic, and Radioactive Waste Management*, **10**(**4**): 16–225. https://doi.org/10.1061/(ASCE)1090-025X(2006)10:4(216).

Goosey, M. and R. Kellner (2003). Recycling technologies for the treatment of end of life printed circuit boards (PCBs). *Circuit World*, **29**(**3**): 33–37. https://doi.org/10.1108/03056120310460801.

Gu, W., Bai, J., Yuan, W., Ma, E., Zhang, C. and J. Wang (2019). Pollution analysis of soil polycyclic aromatic hydrocarbons from informal electronic waste dismantling areas in Xinqiao, China. *Waste Management & Research*, **37**(**4**): 394–401. https://doi.org/10.1177/0734242X19826369.

Islam, M.N., Park, M., Jo, Y.T., Nguyen, X.P., Park, S.S., Chung, S.Y. and J.H. Park (2017). Distribution, sources, and toxicity assessment of polycyclic aromatic hydrocarbons in surface soils of the Gwangju City, Korea. *Journal of Geochemical Exploration*, **180**: 52–60. https://doi.org/10.1016/j.gexplo.2017.06.009.

Kamunda, C., Mathuthu, M. and M. Madhuku (2016). Health risk assessment of heavy metals in soils from Witwatersrand Gold Mining Basin, South Africa. *International Journal of Environmental Research and Public Health*, **13**(**7**): 663. https://doi.org/10.3390/ijerph13070663.

Kumar, S., Negi, S. and P. Maiti (2017). Biological and analytical techniques used for detection of polyaromatic hydrocarbons. *Environmental Science and Pollution Research*, **24**(**33**): 25810–25827. https://doi.org/10.1007/s11356-017-0415-2.

Leung, A.O., Cheung, K.C. and M.H. Wong (2015). Spatial distribution of polycyclic aromatic hydrocarbons in soil, sediment, and combusted residue at an e-waste processing site in southeast China. *Environmental Science and Pollution Research*, **22**(**12**): 8786–8801. https://doi.org/10.1007/s11356-013-1465-8.

Liu, R., Ma, S., Yu, Y., Li, G., Yu, Y. and T. An (2020). Field study of PAHs with their derivatives emitted from e-waste dismantling processes and their comprehensive human exposure implications. *Environment International*, **144**: 106059. https://doi.org/10.1016/j.envint.2020.106059.

Luo, P., Bao, L.J., Li, S.M. and E.Y. Zeng (2015). Size-dependent distribution and inhalation cancer risk of particle-bound polycyclic aromatic hydrocarbons at a typical e-waste recycling and an urban site. *Environmental Pollution*, **200**: 10–15. https://doi.org/10.1016/j.envpol.2015.02.007.

Manoli, E. and C. Samara (1999). Polycyclic aromatic hydrocarbons in natural waters: sources, occurrence and analysis. *TrAC Trends in Analytical Chemistry*, **18**(**6**): 417–428. https://doi.org/10.1016/S0165-9936(99)00111-9.

Miura, K., Shimada, K., Sugiyama, T., Sato, K., Takami, A., Chan, C.K., Kim, I.S., Kim, Y.P., Lin, N.H. and Hatakeyama, S., 2019. Seasonal and annual changes in PAH concentrations in a remote site in the Pacific Ocean. Scientific Reports, 9(1), 1–10. https://doi.org/10.1038/s41598-019-47409-9.

Moeckel, C., Breivik, K., Nøst, T.H., Sankoh, A., Jones, K.C. and A. Sweetman (2020). Soil pollution at a major West African E-waste recycling site: Contamination pathways and implications for potential mitigation strategies. *Environment International*, **137**: 105563. https://doi.org/10.1016/j.envint.2020.105563.

Mojiri, A., Zhou, J.L., Ohashi, A., Ozaki, N. and T. Kindaichi (2019). Comprehensive review of polycyclic aromatic hydrocarbons in water sources, their effects and treatments. *Science of the Total Environment*, **696**: 133971. https://doi.org/10.1016/j.scitotenv.2019.133971.

Nasher, E., Heng, L.Y., Zakaria, Z. and S. Surif (2013). Concentrations and sources of polycyclic aromatic hydrocarbons in the seawater around Langkawi Island, Malaysia. *Journal of Chemistry*, **2013**: 975781. https://doi.org/10.1155/2013/975781.

Nishimura, C., Horii, Y., Tanaka, S., Asante, K.A., Ballesteros Jr, F., Viet, P.H., Itai, T., Takigami, H., Tanabe, S. and T. Fujimori (2017). Occurrence, profiles, and toxic equivalents of chlorinated and brominated polycyclic aromatic hydrocarbons in E-waste open burning soils. *Environmental Pollution*, **225**: 252–260. https://doi.org/10.1016/j.envpol.2016.10.088.

Robinson, B.H. (2009). E-waste: An assessment of global production and environmental impacts. *Science of the Total Environment*, **408(2)**: 183–191. https://doi.org/10.1016/j.scitotenv.2009.09.044.

Sankhla, M.S., Kumari, M., Nandan, M., Mohril, S., Singh, G.P., Chaturvedi, B. and R. Kumar (2016). Effect of electronic waste on environmental & human health-A review. *IOSR Journal of Environmental Science, Toxicology and Food Technology*, **10(9)**: 98–104. https://doi.org/10.1016/j.sbspro.2012.03.305.

Sawyer, C.N., McCarty, P.L. and G.F. Parkin (2003). Chemistry for environmental engineering and science. McGraw-Hill. New York.

Shen, H., Huang, Y., Wang, R., Zhu, D., Li, W., Shen, G., Wang, B., Zhang, Y., Chen, Y., Lu, Y. and H. Chen (2013). Global atmospheric emissions of polycyclic aromatic hydrocarbons from 1960 to 2008 and future predictions. *Environmental Science & Technology*, **47(12)**: 6415–6424. https://doi.org/10.1021/es400857z.Global.

Shi, J., Zheng, G.J.S., Wong, M.H., Liang, H., Li, Y., Wu, Y., Li, P. and W. Liu (2016). Health risks of polycyclic aromatic hydrocarbons via fish consumption in Haimen Bay (China), downstream of an e-waste recycling site (Guiyu). *Environmental Research*, **147**: 233–240. https://doi.org/10.1016/j.envres.2016.01.036.

Sthiannopkao, S. and M.H. Wong (2013). Handling e-waste in developed and developing countries: Initiatives, practices, and consequences. *Science of the Total Environment*, **463**: 1147–1153. https://doi.org/10.1016/j.scitotenv.2012.06.088.

Sun, Z., Xiao, Y., Agterhuis, H., Sietsma, J. and Y. Yang (2016). Recycling of metals from urban mines–a strategic evaluation. *Journal of Cleaner Production*, **112**: 2977–2987. https://doi.org/10.1016/j.jclepro.2015.10.116.

Tang, X., Shen, C., Shi, D., Cheema, S.A., Khan, M.I., Zhang, C. and Y. Chen (2010). Heavy metal and persistent organic compound contamination in soil from Wenling: An emerging e-waste recycling city in Taizhou area, China. *Journal of Hazardous Materials*, **173(1–3)**: 653–660. https://doi.org/10.1016/j.jhazmat.2009.08.134.

USEPA (2004). Risk Assessment Guidance for Super Fund Volume I: Human Health Evaluation Manual (Part E, Supplemental Guidance for Dermal Risk Assessment). Washington, DC, USA.

Wang, W., Simonich, S.L.M., Xue, M., Zhao, J., Zhang, N., Wang, R., Cao, J. and S. Tao (2010). Concentrations, sources and spatial distribution of polycyclic aromatic hydrocarbons in soils from Beijing, Tianjin and surrounding areas, North China. *Environmental Pollution*, **158(5)**: 1245–1251. https://doi.org/10.1016/j.envpol.2010.01.021.

Wang, Y., Tian, Z., Zhu, H., Cheng, Z., Kang, M., Luo, C., Li, J. and G. Zhang (2012). Polycyclic aromatic hydrocarbons (PAHs) in soils and vegetation near an e-waste recycling site in South China: Concentration, distribution, source, and risk assessment. *Science of the Total Environment*, **439**: 187–193. https://doi.org/10.1016/j.scitotenv.2012.08.018.

Wei, L. and Y. Liu (2012). Present status of e-waste disposal and recycling in China. *Procedia Environmental Sciences*, **16**: 506–514. https://doi.org/10.1016/j.proenv.2012.10.070.

Wong, M.H., Wu, S.C., Deng, W.J., Yu, X.Z., Luo, Q., Leung, A.O.W., Wong, C.S.C., Luksemburg, W.J. and A.S. Wong (2007). Export of toxic chemicals–a review of the case of uncontrolled electronic-waste recycling. *Environmental Pollution*, **149(2)**: 131–140. https://doi.org/10.1016/j.envpol.2007.01.044.

Xiang, C.H., Luo, X.J., Chen, S.J., Yu, M., Mai, B.X. and E.Y. Zeng (2007). Polybrominated diphenyl ethers in biota and sediments of the Pearl River Estuary, South China. *Environmental Toxicology and Chemistry: An International Journal*, **26**(**4**): 616–623. https://doi.org/10.1897/06-329R.1.

Xu, P., Tao, B., Ye, Z., Zhao, H., Ren, Y., Zhang, T., Huang, Y. and J. Chen (2016). Polycyclic aromatic hydrocarbon concentrations, compositions, sources, and associated carcinogenic risks to humans in farmland soils and riverine sediments from Guiyu, China. *Journal of Environmental Sciences*, **48**: 102–111. https://doi.org/10.1016/j.jes.2015.11.035.

Yang, T., Cheng, H., Wang, H., Drews, M., Li, S., Huang, W., Zhou, H., Chen, C.M. and X. Diao (2019). Comparative study of polycyclic aromatic hydrocarbons (PAHs) and heavy metals (HMs) in corals, surrounding sediments and surface water at the Dazhou Island, China. *Chemosphere*, **218**: 157–168. https://doi.org/10.1016/j.chemosphere.2018.11.063.

Yu, X.Z., Gao, Y., Wu, S.C., Zhang, H.B., Cheung, K.C. and M.H. Wong (2006). Distribution of polycyclic aromatic hydrocarbons in soils at Guiyu area of China, affected by recycling of electronic waste using primitive technologies. *Chemosphere*, **65**(**9**): 1500–1509. https://doi.org/10.1016/j.chemosphere.2006.04.006.

Zhang, W.H., Ying-Xin, W.U. and M.O. Simonnot (2012). Soil contamination due to e-waste disposal and recycling activities: A review with special focus on China. *Pedosphere*, **22**(**4**): 434–455. https://doi.org/10.1016/S1002-0160(12)60030-7.

Chapter 7
Bioluminescence Inhibition Assay for Toxicity Measurement of PAHs and Heterocyclic PAHs

Prasenjit Ghosh and Suparna Mukherji

1 Introduction

Industrialization and urbanization have caused the entry of new chemicals into the environment. Most of the chemicals are discharged from different industrial processes in the form of components of complex mixtures and their toxicities cannot be determined by conventional water quality parameters, such as chemical oxygen demand (COD) and biochemical oxygen demand (BOD). Since the compounds are discharged as complex mixtures, it is very difficult to predict the toxicity of mixtures even when the toxicity of the pure components is known. Hence, the selection of an appropriate bioassay is necessary. The most suitable bioassay may depend upon the sample characteristics, associated cost and simplicity (Parvez et al. 2006).

Polynuclear aromatic hydrocarbons (PAHs) are ubiquitous organic pollutants whose natural sources include forest fires, volcanic eruptions, crude oils, plant alkaloids and soil organic matter and anthropogenic sources include oil spills, fossil fuel combustion, food cooked at high temperatures and wastewater generated from various industries (Salam et al. 2017). The U.S. EPA listed 16 PAHs as "priority PAHs" in 1976 and since that time these compounds are regularly monitored for assessment of the risk of oil-contaminated sites (Keith 2015). The emission of PAHs is often associated with the release of heterocyclic PAHs, i.e., N, S and O-substituted PAHs. Hence these compounds often coexist with the PAH compounds in various contaminated sites. The pollution potency of the heterocyclic PAHs was identified between the 1970s and 1980s (Bleeker et al. 2002). These compounds have been detected at concentrations ranging from 0.2 to 61,000 ng/g in various contaminated soil and sediment sites, 6 ng/L–3173 mg/L in contaminated groundwater, 0.1 ng/L–84 μg/L in surface water (rivers and oceans) and 10 ng/L–2000 mg/L in various

P. Ghosh · S. Mukherji (✉)
Environmental Science and Engineering Department (ESED), Indian Institute of Technology Bombay, Mumbai 400 076, India
e-mail: mitras@iitb.ac.in

S. Goel (ed.), *Advances in Solid and Hazardous Waste Management*,
https://doi.org/10.1007/978-3-031-49144-3_7

industrial wastewaters (Dyreborg et al. 1996). Although their abundance in the environment is usually much lower relative to PAHs (approximately 10% of the PAHs), they are characterized by higher acute toxicity compared to PAHs and several heterocyclic PAHs also exhibit mutagenicity and carcinogenicity (Salam et al. 2017).

Various physical, chemical, physicochemical, biological and combined treatment technologies have been applied for the remediation of oil-contaminated sites, however, biological treatment processes have been identified as more economical and eco-friendly compared to the other processes. Biological treatment processes often involve the application of bacteria, fungus, algae, yeast and plants for the remediation of contaminated sites. However, the PAHs and heterocyclics often inhibit the bioremediation process. Moreover, the low molecular weight PAHs inhibit the degradation of high molecular weight heterocyclics and vice versa (Meyer and Steinhart 2000). Hence, it is very important to determine the toxicity and also determine variation in toxicity during biodegradation in various contaminated matrices.

Bioassays involve the use of animals or plants (in vivo) or tissues or cells (in vitro) in order to determine the biological effects of a substance on living organisms. The application of bioassays is essential for monitoring various pollutants in contaminated sites. A bioassay can also be used to determine the toxic concentration of a substance. Hence, both qualitative and quantitative analyses of the toxic nature of a substance can be done using bioassays (Parvez et al. 2006). In-vivo assays are expensive and labour-intensive. Microbioassays using bacteria or enzymes are being increasingly used for the measurements of toxicity of chemicals in the environment due to their low cost, high sensitivity, high sample throughput, modest laboratory instrument and space requirements, low sample volume, portability and reproducible responses (Parvez 2008). Bacterial toxicity tests are based on the measurement of bioluminescence, growth, motility and oxygen uptake. The permeability of bacterial cells to environmental toxicants makes the bacterial assays more sensitive (Bitton and Koopman 1992).

Some of the commercially available microbioassays are Microtox (bioluminescence), Polytox (respiration), Toxi-Chromotest (enzyme biosynthesis), SOS-Chromotest (genotoxicity of chemicals), MetPAD (enzyme activity) and ECHA Biocide Monitor (dehydrogenase activity). A standard lyophilized bacterial culture is used in all these bacterial bioassays (Bitton and Koopman 1992). El-Alawi et al. (2002) measured short- and long-term toxicities of twelve PAHs using a luminescent bacteria, *Vibrio fischeri* NRRL B-11177. Mersch-Sundermann et al. (2013) measured the toxicities of 32 PAHs using SOS Chromotest which is a genotoxicity assay using *E. coli* PQ3. Bispo et al. (1999) measured toxicity (acute and chronic) and genotoxicity of PAH-polluted soils in an industrial area of France. Acute toxicity was measured by Microtoxtest using *Vibrio fischeri*, *Daphnia magna* (crustaceans) and *Thamnocephalu splatyurus* (freshwater crustacean) and chronic toxicity were measured using *Pseudokirchneriella subcapitata* (fresh water microalgae). Genotoxicity was measured using the Mutatox test based on *Vibrio fischeri*. Backhaus et al. (2000) used a bioluminescence inhibition assay for measuring the toxicity of quinolones.

Use of luminescent bacterial cultures as bioindicators started in the 1950s (Scheerer et al. 2006). The use of *Vibrio fischeri* bioluminescence assays for toxicity measurement of polluted water, soil, and sediment samples had increased over time (Boluda et al. 2002). *Vibrio fischeri* is a bioluminescent proteobacterium that is found mostly in marine environments and most commonly found in symbiont relations with other marine animals such as squids and fishes. They may also exist as free living cells. They are gram-negative and rod shaped bacteria that produce bioluminescence. Bacterial bioluminescence is inhibited when the cells are exposed to toxicant(s) over a fixed time period. ISO 11348 describes three different methods for determining inhibition in luminescence using freshly prepared, liquid dried and freeze driedform of the bioluminescent bacterial strain, *Vibrio fischeri* NRRL B-11177. It is now renamed as *Aliivibrio fischeri.* The bacterial strain (in liquid and freeze-dried forms) has been widely used in several commercial toxicity test kits, such as Microtox, LUMIStox, ToxAlert, etc. However, they are very expensive and have a limited life span. Toxicity studies using freshly prepared *Vibrio fischeri* NRRL B-11177 as described in ISO 11348-1:2007 (BS EN ISO 2007) offer significant cost savings and are not affected by limited self-life (Parvez 2008).

The first objective of the current study was to determine the dose-response profile of some selected PAHs and heterocyclic PAHs using the bioluminescent bacterial strain, *Vibrio fischeri* NRRL B-11177. The second objective was to develop correlations among experimental EC_{50} (half maximal effective concentration), EC_{50} of the compounds based on other microbes and K_{ow} (octanol–water partition coefficient) of the compounds. The third and final objective was to determine the acute toxicity profile during degradation of a nitrogen heterocyclic PAH, carbazole by a bacterial strain, *Pseudomonas aeruginosa* RS1, which is capable of efficiently degrading a wide range of PAHs and some heterocyclic PAHs.

2 Materials and Methods

2.1 Chemicals

Crabazole ($\geq$95%) and fluorene (98%) were purchased from Sigma Aldrich Chemicals Pvt. Ltd., St. Louis, USA. Pyridine (99%), thiophene (99%), benzo(b)thiophene (98%) and dibenzothiophene (98 + %) were purchased from Thermo Fisher Scientific India Pvt Ltd. Benzene ($\geq$99.5%), quinoline (97%) and naphthalene (99%) were obtained from Merck, India Ltd., S D Fine-Chem Limited, Mumbai and Loba-Chemie, India Ltd., respectively. HPLC grade acetonitrile was procured from Merck, India. Chemicals for preparing media for performing the bioluminescence inhibition tests were purchased from Merck, India Ltd., Ranbaxy, India Ltd., Thomas Baker (Chemicals) Ltd., Mumbai, India, LobaChemie, India Ltd. and Finar Ltd., Gujarat, India.

2.2 Bacterial Culture Used for Carbazole Degradation Studies

Pseudomonas aeruginosa RS1(MTCC 25391), originally isolated from an oil refinery sludge was used for conducting the biodegradation studies on carbazole (100 mg/L initial concentration). The bacterial strain was earlier reported to degrade carbazole, fluorene and dibenzothiophene in mineral media as well as in refinery wastewater (Ghosh and Mukherji 2020a, b). The methods of inoculum preparation and setting up the biodegradation experiments are discussed elsewhere (Ghosh and Mukherji 2018, 2020b).

2.3 Toxicity Testing Using Aliivibrio Fischeri NRRL B-11177

Bioluminescence inhibition exhibited by the strain upon exposure to a toxic chemical for a fixed period indicates non-specific acute toxicity. Lyophilized cells of *Vibrio fischeri* NRRL B-11177 were purchased from ABOATOX, Finland and used for developing dose-response relationships of the compounds. The strain was also used for measuring the toxicity of cell free broths during the degradation of carbazole. The strain was grown on ISO-recommended liquid broth for pre- and main cultures at 20 ± 1 °C and 180 rpm and stock suspension and test suspension were prepared following the standard method described in BS EN ISO (2007). The test suspension was used for conducting the luminescence inhibition studies. pH of the undiluted test sample was adjusted to 7.0 ± 0.2 using 2N HCl/1 M NaOH solution. Different dilution series of the pH adjusted sample were prepared using 2% NaCl as diluent. 2% NaCl solution was used as the control. Test suspension of 0.5 ml was dispensed in all the luminometer tubes and their I_0 values were measured. Each of the reference samples (0.5 ml), undiluted samples, diluted samples and controls were thereafter added to the luminometer tubes and were maintained at 15 ± 1 °C for at least 30 min. All the samples and controls were measured in duplicate. In order to maintain equal contact time for all the test samples and controls, luminescence intensities of the samples were measured at 10 s intervals after a contact time of 15 and 30 min. Percent inhibition (H_t in %) was determined and was used for plotting the dose-response profiles for determining the acute toxicity of the compounds.

2.4 Optimization of Bioluminescence Intensity of Aliivibrio Fischeri NRRL B-11177

In order to prepare the test suspension, the bacterial cultures should be harvested when they reach maximum luminescence intensity. The optimum time for harvesting the bacterial cells was determined by culturing the strain in duplicate 500 ml Erlenmeyer flasks. Freshly grown *Vibrio fischeri* culture (2.5 ml) was added in 2 Erlenmeyer flasks containing 100 ml of the growth media. After every 4 h, 3 ml of the culture was aseptically taken from each of the Erlenmeyer flasks and absorbance and luminescence values of the samples were measured using a double beam spectrophotometer at 578 nm wavelength and a luminometer, respectively. Measurement was continued until a decrease in luminescence was observed. The volume of the culture was replenished by adding 3 ml of bacterial growth media in the culture flasks at every sampling time. Subculturing was done by adding 2.5 ml of the grown culture to another set of 2 Erlenmeyer flasks containing 100 ml of growth media. Samples were taken from both of the Erlenmeyer flasks and the same measurement procedure for absorbance and luminescence was repeated. This was done to validate the optimum time for harvesting the culture for the bioluminescence inhibition studies.

2.5 *Measurement of Threshold Toxic Concentration of Solvent*

Since some of the compounds are sparingly soluble in aqueous solution (such as carbazole, fluorene and dibenzothiophene), an organic solvent was used to dissolve these compounds prior to toxicity measurements. Acetonitrile (ACN) was chosen as the solvent as it is water miscible and could solubilize various compounds of interest. In order to determine the maximum safe concentration of acetonitrile in the test samples that would not lead to significant toxic effects, the threshold concentration of acetonitrile was measured using the bioluminescence inhibition assay. Different dilutions of acetonitrile were prepared using 2% NaCl as a diluent. Initially, samples containing acetonitrile ranging from 100% to 0.1% were prepared using 2% NaCl as a diluent. The density of acetonitrile in its purest form is 0.78 kg/L. Based on the result obtained, the dilution series was narrowed down to (20–1%) ACN, in order to determine the safe % of ACN in the samples that would not cause a significant toxic effect.

EC_5 (concentration of ACN at which 5% of maximal toxic response[1] is seen) at 5, 15 and 30 min exposure times were compared for both experiments to determine the threshold value. The stock solution of carbazole, fluorene, benzothiophene and dibenzothiophene was prepared in ACN at various concentrations and the samples were prepared by diluting with 2% NaCl, such that % of ACN was less than its toxic threshold value. Stock solutions of the other substrates (pyridine, quinoline, naphthalene, thiophene and benzene) were prepared in a 2% NaCl solution and subsequently diluted using the same.

[1] Toxic response refers to decrease in luminescence of Vibrio fischeri NRRL B-11177.

2.6 Dose-Response Relationships of Various PAHs and Heterocyclic PAHs

Dose-response relationships were developed for determining the acute toxicity of a total of 9 compounds containing one monoaromatic (benzene), two PAHs (naphthalene and fluorene), and six nitrogen and sulphur heterocyclics (pyridine, quinoline, carbazole, thiophene, benzothiophene and dibenzothiophene) for comparing their acute toxicity. Static toxicity assay was performed as per BS EN ISO 11348-1: 2007 over an exposure time of 30 min, using the freshly prepared cells of *Vibrio fischeri* NRRL B-11177. The bacterial cells were harvested after 20–24 h of incubation since the culture showed maximum luminescence within this time period. The threshold concentration of acetonitrile showing less than 5% bioluminescence inhibition value was found to be 5.6%. Hence, the stock solution of the solid compounds (carbazole, fluorine, benzothiophene and dibenzothiophene) was prepared in 5% acetonitrile to avoid any interference due to the solvent. Toxicity comparisons were made based on the EC_{50} (measured after 15 and 30 min exposure time) values of the compounds, i.e., the concentration of a particular compound causing 50% light reduction in *Vibrio fischeri* NRRL B-11177.

2.7 Determination of Acute Toxicity During Carbazole Degradation

The carbazole degradation experiment was set up at 100 mg/L initial concentration over a period of 14 days and aliquots were collected at 2 days intervals. The samples were filtered under aseptic conditions using 0.2 μm nylon membrane filters and the filtrates were stored in sterile serum bottles. The luminescence inhibition test was conducted using filtrates.

3 Results and Discussion

3.1 Determination of Acute Toxicity of Various PAHs and Heterocyclic PAHs by Luminescence Inhibition Test

3.1.1 Optimization of Bioluminescence Intensity of Vibrio Fischeri

Vibrio fischeri culture should be harvested when it reaches maximum luminescence intensity. The time for reaching maximum luminescence intensity was determined through two successive subculturing stages. The luminescence of controls and absorbance of both the controls and the cultures was measured at 578 nm wavelength. The time for reaching the maximum luminescence value in the two consecutive subculturing stages was 20 and 24 h, respectively. Hence, the *Vibrio fischeri* cultures were harvested after 20–24 h of incubation and the cell pellets obtained were used for preparing stock suspensions for the toxicity studies.

3.1.2 Measurement of Threshold Toxic Concentration of Solvent

The objective of this experiment was to determine the maximum concentration of acetonitrile below which is not toxic to the *Vibrio fischeri* cells. The experiment was initially done in the 100–0.1% dilution range and subsequently repeated over the 20–1% dilution range. For the 100–0.1% dilution series, EC_5 was 8.1% acetonitrile and for the 20–1% dilution series, EC_5 was 5.6% acetonitrile. Hence, 5% acetonitrile was used for solubilizing the sparing organics while avoiding any significant toxic effect due to acetonitrile.

3.1.3 Dose Response Relationships of PAHs and Heterocyclic PAHs

Using the *Vibrio fischeri* bioluminescence inhibition test, EC_{50} of one monoaromatic, two PAHs and six heterocyclics (three PANH and three PASH compounds) were determined after 15 min and 30 min exposure duration and the values are listed in Table 1. Figures 1, 2 and 3 depict the dose-response profiles of the compounds. Based on the EC_{50} values (mg/L) of the compounds it can be concluded that an increase in the number of rings from one to three within a particular group of compounds (PANH and PASH) increased the acute toxicity. Thus, for PANHs, carbazole (CBZ) was the most toxic followed by quinolone (Q) and pyridine (P). For PASHs, dibenzothiophene (DBT) was the most toxic followed by benzothiophene (BT) and thiophene (Th). Fluorene (FLU) was the most toxic followed by naphthalene (N) and benzene (B). When one carbon atom in a ring was replaced by a nitrogen or sulfur atom, toxicity of the resulting compound decreased, except for the fluorene and dibenzothiophene pair. Among the one-ring compounds, benzene was the most toxic followed by thiophene and pyridine. Among the two-ring compounds, naphthalene was the most toxic followed by quinoline and benzothiophene. Dibenzothiophene was the most toxic compound followed by fluorene and carbazole among the three-ring compounds. Dibenzothiophene and pyridine were the most toxic and the least toxic compounds, respectively, among all the compounds. The compounds in descending order of non-specific acute toxicity are DBT > FLU > N > CBZ > Q > BT > B > Th > P.

Figures 4 and 5a–d represent the correlations between experimental EC_{50} of the compounds and their K_{ow} values, experimental EC_{50} of the compounds and their literature-reported values with respect to *Vibrio fischeri*, experimental EC_{50} of CBZ, DBT and Q and EC_{50} with respect to *Daphnia magna*, experimental EC_{50} of Th, BT, DBT and Q and EC_{50} with respect to *Pseudomonas fluorescence* and experimental EC_{50} of P, BT, CBZ and Q and LC_{50} with respect to *Danio rerio*, respectively. K_{ow}

Table 1 Summary of EC_{50} values of various PAHs and heterocyclic PAHs

Compounds	EC_{50} (mg/L)		[a]Aqueous solubility at 25 °C (mg/L)
	After 15 min	After 30 min	
Benzene (B)	395.65	97.91	1790
Naphthalene (N)	1.53	0.89	31
Fluorene (FLU)	0.82	0.63	1.69
Pyridine (P)	459.74	375.76	Miscible
Quinoline (Q)	6.44	6.96	6110
Carbazole (CBZ)	2.13	1.63	1.8
Thiophene (Th)	399.03	166.27	3010
Benzothiophene (BT)	51.79	45.35	130
Dibenzothiophene (DBT)	0.72	0.51	1.47

[a] *Source* URL 1

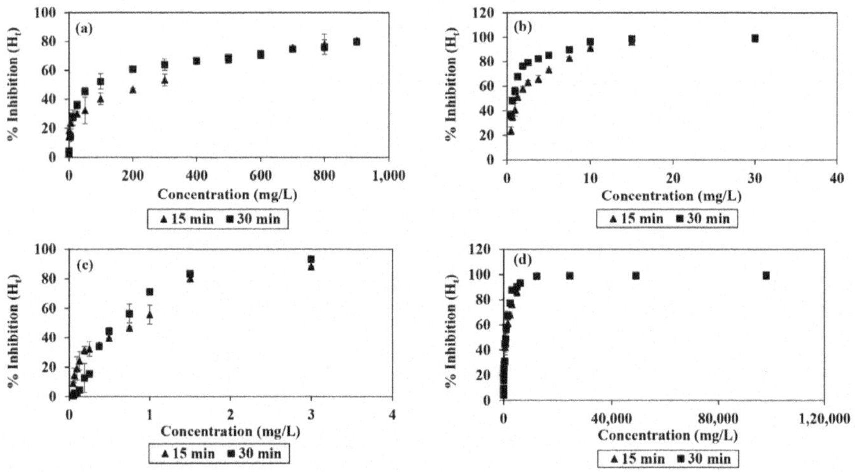

Fig. 1 Dose response profiles of **a** benzene, **b** naphthalene, **c** fluorine and **d** pyridine after 15 and 30 min exposure

and EC_{50} values of the monoaromatic benzene and two PAHs, naphthalene and fluorene were taken from various sources (Docherty and Kulpa 2005; Lee et al. 2013). Log(EC_{50}) was inversely correlated with Log(K_{ow}) with R^2 of 0.78 (Fig. 4). Correlation between experimental EC_{50} and EC_{50} values obtained from the literature for all the nine compounds was good except for BT and FLU (Fig. 5a). A good correlation ($R^2 = 0.94$) was observed between experimental EC_{50} values and EC_{50} of *Daphnia magna* (obtained from literature) for DBT, CBZ and Q (Fig. 5b). Figure 5c, d show the correlations between experimental EC_{50} values and EC_{50} of *Pseudomonas fluorescence* and LC_{50} of *Danio rerio* (obtained from the literature) for various compounds.

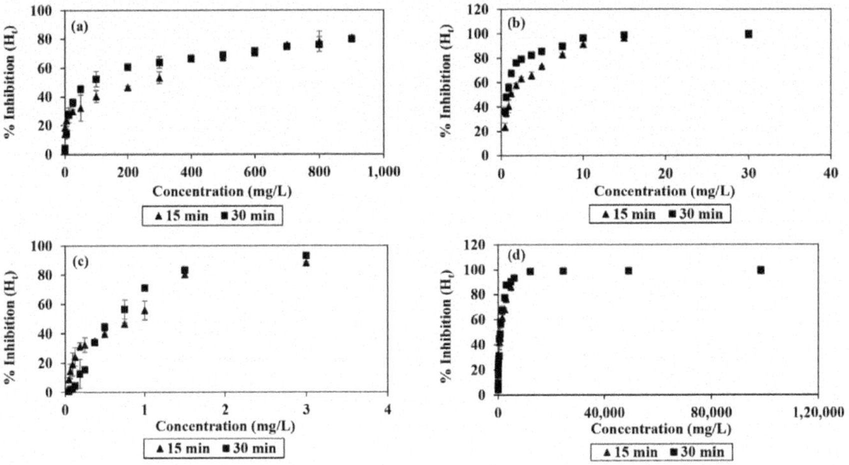

Fig. 2 Dose response profiles of **a** quinoline, **b** carbazole, **c** thiophene and **d** benzothiophene after 15 and 30 min exposure

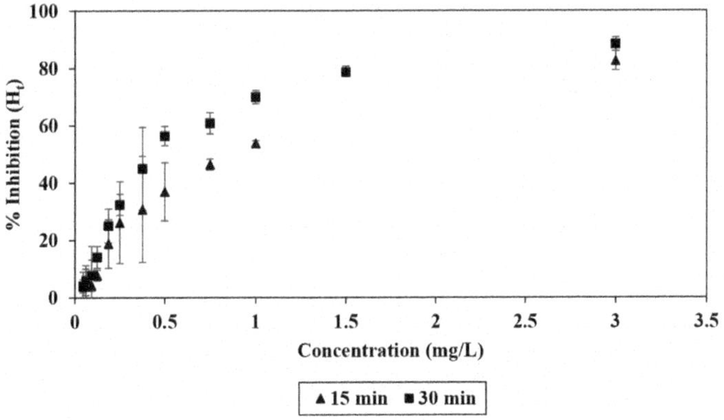

Fig. 3 Dose response profiles of dibenzothiophene after 15 and 30 min exposure

3.2 *Determination of Acute Toxicity During Carbazole Degradation*

Figure 6 depicts the inhibition profile (H_t, %) of the broth samples collected during various time points of carbazole (100 mg/L) degradation towards luminescence of *Vibrio fischeri* culture. H_t values were obtained after 15 and 30 min exposure times of mixing the test suspension and the samples. In both cases, H_t value increased till 10th day, which subsequently was slightly reduced till the 14th day. Maximum inhibition of 63 ± 6% was observed after 30 min incubation for the samples collected on the 10th day. No significant change was observed for the controls over 14 days.

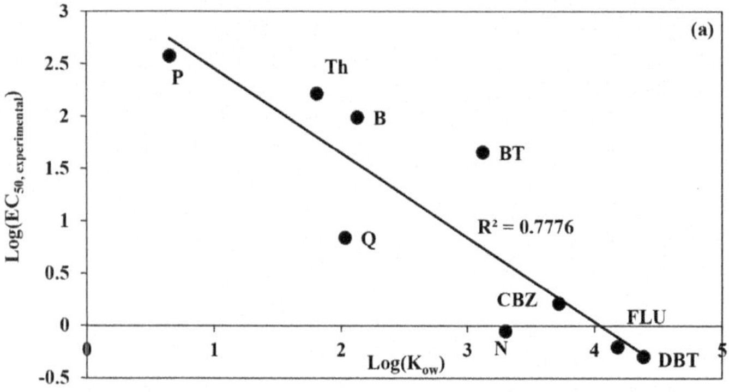

Fig. 4 Correlation between experimental EC_{50} and K_{ow} of various PAHs and heterocyclic PAHs

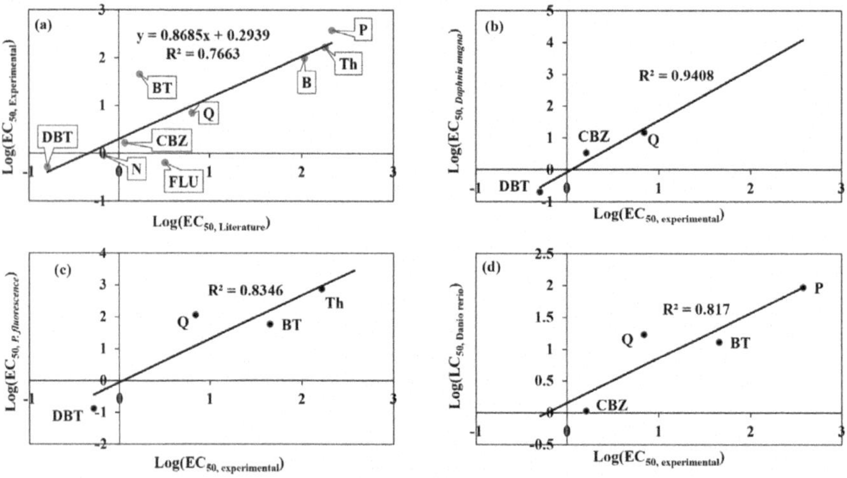

Fig. 5 Correlation between experimental EC_{50} and EC_{50} of various PAHs and heterocyclic PAHs observed in **a** literature for Vibriofischeri, **b** Daphnia magna, **c** *Pseudomonas fluorescence* and **d** Danio rerio after 30 min contact time. *Note* B: benzene, N: naphthalene, FLU: fluorene, P: pyridine, Q: quinoline, CBZ: carbazole, Th: thiophene, BT: benzothiophene and DBT: dibenzothiophene

4 Conclusions

EC_{50} (15 and 30 min) values of the compounds imply high toxic effects of naphthalene, fluorene, quinoline, carbazole and dibenzothiophene. It can be noted that the sparingly soluble compounds, fluorene, carbazole and dibenzothiophene are toxic even at their aqueous solubility levels. Thus, they pose a risk to aquatic organisms. Correlation between the experimentally observed and literature-reported EC_{50} values for one monoaromatic, two PAHs and six heterocyclics indicates a good

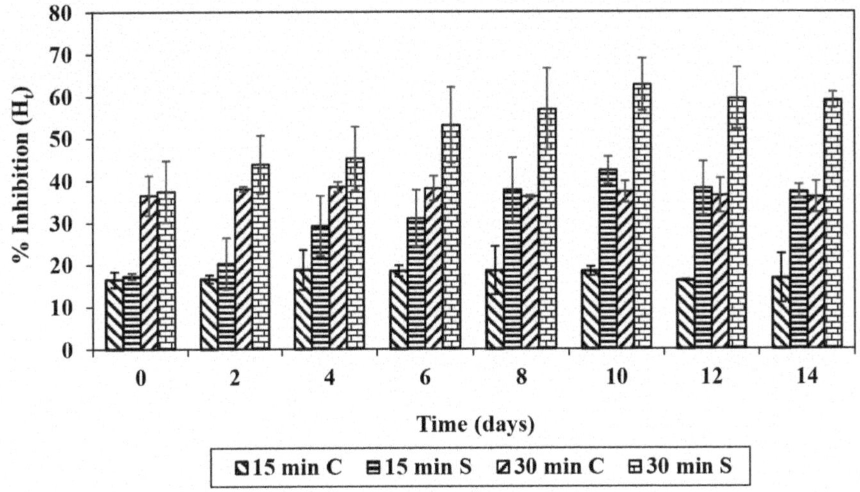

Fig. 6 Luminescence inhibition profiles of broth samples collected during carbazole degradation. *Note* C and S represent control and sample, respectively

match between them. A good correlation was also observed between experimentally observed EC_{50} values and EC_{50}/LC_{50} values of some compounds observed in bioassays with *Daphnia magna*, *Pseudomonas fluorescence* and *Danio rerio*. Acute toxicity of samples measured during carbazole degradation revealed that maximum luminescence inhibition was observed on the 10th day of incubation. Accumulation of toxic metabolites enhanced the toxicity during the biodegradation of 100 mg/L CBZ.

References

Backhaus, T., Scholze, M. and L.H. Grimme (2000). The single substance and mixture toxicity of quinolones to the bioluminescent bacterium *Vibrio fischeri*. Aquat. Toxicol., **49**: 49–61.

Bispo, A., Jourdain, M.J. and M. Jauzein (1999). Toxicity and genotoxicity of industrial soils polluted by polycyclic aromatic hydrocarbons (PAHs). *Org. Geochem.* **30**: 947–952.

Bitton, G. and B. Koopman (1992). Bacterial and enzymatic bioassays for toxicity testing in the environment. *Rev. Environ. Contam. Toxicol.*, **125**: 1–22.

Bleeker, E.A.J., Wiegman, S., Voogt, P. De, Kraak, M., Leslie, H.A., Haas, E. De and W. Admiraal (2002). Toxicity of azaarenes. *Rev. Env. Contam. Toxicol.*, **173**: 39–83.

Boluda, R., Quintanilla, J.F., Bonilla, J.A., Sáez, E. and M. Gamón (2002). Application of the Microtox® test and pollution indices to the study of water toxicity in the Albufera Natural Park (Valencia, Spain). *Chemosphere*, **46**: 355–369.

BS EN ISO (2007). BS EN ISO 11348-1:2007—Water quality—Determination of the inhibitory effect of water samples on the light emission of *Vibrio fischeri* (Luminescent bacteria test), 2007.

Docherty, K.M. and C.F. Kulpa Jr. (2005). Toxicity and antimicrobial activity of imidazolium and pyridinium ionic liquids. *Green Chem.*, **7**: 185–189.

Dyreborg, S., Arvin, E. and K. Broholm (1996). Effects of creosote compounds on the aerobic bio-degradation of benzene. *Biodegradation*, **7**: 191–201. https://doi.org/10.1007/BF00058179.

El-Alawi, Y.S., McConkey, B.J., Dixon, D.G. and B.M. Greenberg (2002). Measurement of short-and long-term toxicity of polycyclic aromatic hydrocarbons using luminescent bacteria. *Ecotoxicol. Environ. Saf.*, **51**: 12–21.

Ghosh, P. and S. Mukherji (2018). Optimization of media composition for enhancing carbazole degradation by *Pseudomonas aeruginosa* RS1, *J. Environ. Chem. Eng.* **6**: 2881–2891. https://doi.org/10.1016/j.jece.2018.04.043.

Ghosh, P. and S. Mukherji (2020a). Degradation of carbazole, fluorene, dibenzothiophene and their mixture by *P. aeruginosa* RS1 in petroleum refinery wastewater. *J. Water Process Eng.*, **37**: 101454.

Ghosh, P. and S. Mukherji (2020b). Modeling growth kinetics and carbazole degradation kinetics of a *Pseudomonas aeruginosa* strain isolated from refinery sludge and uptake considerations during growth on carbazole. *Sci. Total Environ.*, **738**: 140277.

Keith, L.H. (2015). The source of US EPA's sixteen PAH priority pollutant. *Polycycl. Aromat. Compd.*, **35**: 147–160.

Lee, S.Y., Kang, H.J. and J.H. Kwon (2013). Toxicity cut off of aromatic hydrocarbons for luminescence inhibition of *Vibrio fischeri*. *Ecotoxicol. Environ. Saf*. **94**: 116–122.

Mersch-Sundermann, V., Mochayedi, S. and S. Kevekordes (2013). Genotoxicity of polycyclic aromatic hydrocarbons in *Escherichia coli* PQ37. *Adv. Mol. Toxicol.*, **278**: 1–9.

Meyer, S. and H. Steinhart (2000). Effects of heterocyclic PAHs (N, S, O) on the biodegradation of typical tar oil PAHs in a soil/compost mixture. *Chemosphere*, **40**: 359–367.

Parvez, S., Venkataraman, C. and S. Mukherji (2006). A review on advantages of implementing luminescence inhibition test (*Vibrio fischeri*) for acute toxicity prediction of chemicals, *Environ. Int.* **32**: 265–268.

Parvez, S. (2008). Bioluminescence Inhibition Based Toxicity Assessment of Hydrophobic Organic Compounds: Component Contribution to Mixture Toxicity, Ph.D. Thesis, CESE, IIT Bombay.

Salam, L.B., Ilori, M.O. and O.O. Amund (2017). Properties, environmental fate and biodegradation of carbazole. *3 Biotech*, **7**: 111.

Scheerer, S., Gomez, F. and D. Lloyd (2006). Bioluminescence of *Vibrio fischeri* in continuous culture: Optimal conditions for stability and intensity of photoemission. *J. Microbiol. Methods*, **67**: 321–329. URL 1: https://pubchem.ncbi.nlm.nih.gov/.

Chapter 8
Method Development for Identification and Quantification of PAHs and Heterocyclic PAHs and Their Degradation Intermediates

Prasenjit Ghosh and Suparna Mukherji

1 Introduction

Polynuclear aromatic hydrocarbons (PAHs) are organic pollutants having two or more fused aromatic rings. Discovery of the carcinogen, benzo(a)pyrene in 1930 triggered interest within the research community about PAHs. PAHs are released into the environment during oil spills, fossil fuel combustion, cooking food at high temperatures and wastewater discharges from various industries (such as dye, plastic and pharmaceuticals, insecticides, drugs and polymers and oil refining industries) (Ghosh and Mukherji 2020). Several PAHs have been reported to be toxic, mutagenic, genotoxic, and some are even carcinogenic (Haritash and Kaushik 2009). In the year 1976, the U.S. EPA identified 16 PAHs as "priority pollutants" and subsequently these compounds have been regularly monitored in oil-contaminated sites for risk assessment (Keith 2015).

However, emissions of PAHs are often associated with the release of heterocyclic PAHs and their pollution potency was identified between the 1970s and 1980s (Bleeker et al. 2002). Unlike PAHs and BTEX (benzene, toluene, ethyl benzene and xylene) compounds, heterocyclic PAH compounds are not yet included in standard environmental monitoring programmes, even in developed countries (Mundt and Hollender 2005). Although their environmental abundance is up to 1–10% of the analogues PAHs, they pose significant ecotoxic effects (both acute and chronic toxicity, including mutagenicity and carcinogenicity) to various animals and plants at oil contaminated sites (Salam et al. 2017). Compared to PAHs, heterocyclic PAHs often pose higher toxicity even when they are present in trace amounts. Higher polarity

P. Ghosh · S. Mukherji (✉)
Environmental Science and Engineering Department (ESED), Indian Institute of Technology Bombay, Mumbai 400 076, India
e-mail: mitras@iitb.ac.in

S. Goel (ed.), *Advances in Solid and Hazardous Waste Management*,
https://doi.org/10.1007/978-3-031-49144-3_8

and aqueous solubility of the low molecular weight heterocyclic PAHs enhance their bioavailability and make them highly potent environmental contaminants. Considering the pollution potency of PAH and heterocyclic PAH compounds, it is very important to develop identification and quantification techniques for conducting a risk assessment of oil-contaminated sites. These compounds are often degraded by various natural processes into low molecular weight compounds, often referred to as intermediates or metabolites. In several cases, the presence of some intermediates in contaminated sites implies the presence of the parent PAH and heterocyclic PAH compounds. The toxicity of the intermediates is often more compared to the parent compounds, such that their natural attenuation is adversely affected (Geier et al. 2018). Hence, it is important to develop identification and quantification methods for the metabolites of heterocyclic PAHs and understand the degradation pathways.

PAHs and heterocyclic PAHs are present in the form of different structural isomers in the environment. Gas chromatography (GC-FID and GC-MS) and High Performance Liquid Chromatography (HPLC) techniques are generally used for isomer-specific identification and quantification after extracting the compounds using liquid–liquid extraction or solid phase micro-extraction from environmental samples. Reverse phase liquid chromatography (RPLC) systems with fluorescence detectors are often used for the separation of nitrogen heterocyclic PAHs (PANH) from PAH-PANH mixtures and routine analyses of PANH compounds using acetonitrile-aqueous buffer as the mobile phase (Švábenský et al. 2007). Their metabolites can be identified and quantified using LC with diode array, fluorescence or mass selective detectors after performing liquid or solid phase extraction (Mundt and Hollender 2005). Some other technologies involved are Nuclear Magnetic Resonance Spectroscopy (NMR), Thin Layer Chromatography (TLC), Remote Fiber Spectroscopy (RFS), Size Exclusion Chromatography, Infrared Spectroscopy, Supercritical Fluid Chromatography, Ultraviolet and fluorescence spectroscopy and Isotope ratio mass spectroscopy.

There are several HPLC methods available in the literature on the identification and quantification of various PAHs (Mollea et al. 2005; Akdogan and Pazarlioglu 2011; Mosmeri et al. 2017), PANHs (Lee et al. 1994; Wang et al. 2007; Bai et al. 2010), sulphur heterocyclic PAHs (PASH) (van Afferden et al. 1989; Eaton and Nitterauer 1994; Rivas et al. 2003), oxygen heterocyclic PAHs (PAOH) (Celis et al. 2006) and their mixtures (Chupungars et al. 2009). While acetonitrile and methanol are the most frequently used organic phase (acetonitrile being more widely used), water or water supplemented with phosphate buffer, ammonium acetate buffer or H_2SO_4 (up to 0.12%) have been used as the aqueous phase for constituting the mobile phase. For the identification of individual compounds, the isocratic mode of mobile phase application has been reported at organic phase to aqueous phase proportion varying from 98:2 to 60:40. Method development for compound mixtures often involves the application of gradient mode of mobile phase application. Few literature studies reported HPLC methods for the identification and quantification of PASH compounds individually and in mixtures, compared to the other compound groups. Saha et al. (2010)

developed an HPLC method for the simultaneous identification and quantification of 4 PANH compounds namely, pyridine, 2-picoline, 4-picoline and quinoline from cigarette smoke. About 0.01 M ammonium acetate buffer was used as the aqueous phase and acetonitrile were used as the organic phase, with a 1 ml/min flow rate of the mobile phase (total 9 min run time) in gradient mode. Meyer and Steinhart (2000) reported an HPLC method for the simultaneous identification of PAHs and heterocyclic PAH compounds (PAHs: naphthalene, acenaphthene, fluorene, phenanthrene, anthracene, fluoranthene, pyrene, benzo[a]anthracene and benzo[a]pyrene; PANHs: indole, quinoline, carbazole and acridine; PASHs: benzothiophene, dibenzobenzothiophene; PAOHs: Benzofuran and dibenzofuran). A ternary gradient, comprised of acetonitrile, phosphate buffer and methanol was used as the mobile phase.

There are several studies in the literature on the identification and quantification of PAHs (Byrd et al. 1990; Collina et al. 2005; Lin et al. 2010; Akdogan and Pazarlioglu 2011), PANHs (Aislabie et al. 1990; Rajagopal et al. 1990; Huang et al. 2003), PASHs (Eaton and Nitterauer 1994; Gilbert et al. 1998; Stumpf et al. 1998; Bahrami et al. 2001) and PAOHs (Meyer and Steinhart 2000) and their mixtures (Meyer and Steinhart 2000; Jensen et al. 2003; Flego and Zannoni 2011), using GC, GC-MS and GC-GC-TOF MS. From the literature it can be concluded that the GC oven temperature method contains mostly single ramp for identification of single compounds (except for compounds having a low boiling point in the range of 80–120 °C), whereas, multiple ramps are provided for identification of mixtures. It can also be concluded that the GC oven temperature method for compounds having low boiling points involved much slower ramp rates compared to compounds having higher boiling points (in the range of 200–400 °C).

Several researchers have discussed methods for the identification and quantification of metabolites formed during the degradation of various PAHs and heterocyclic PAHs using HPLC or HR-LCMS (Berthou and Vignier 1986; Grifoll et al. 1995; Bezalel et al. 1996; Hammer et al. 1998; Traulsen et al. 1999) and GC-MS (Eaton and Nitterauer 1994; Gieg et al. 1996; Bahrami et al. 2001; Jianlong et al. 2002; Gai et al. 2008) methods. Metabolite identification is a pivotal step in exploring the degradation pathway(s) of a compound. For analysis of metabolites using GC, the samples are usually derivatized before analysis, however, in some studies, derivatization was not reported prior to GC analysis. The sole purpose of derivatization is to enhance the volatility of the metabolites by changing their polarity.

The current study focuses on developing HPLC, GC-FID and GC × GC-TOFMS methods for the identification and quantification of various PAHs and heterocyclics individually and in mixtures.

2 Materials and Methods

2.1 *Chemicals*

Carbazole (≥95%) and fluorene (98%) were obtained from Sigma Aldrich Chemicals Pvt. Ltd., St. Louis, USA. Pyridine (99%), thiophene (99%), benzo(b)thiophene (98%) and dibenzothiophene (98+%) were purchased from Thermo Fisher Scientific India Pvt Ltd. Benzene (≥99.5%), quinoline (97%) and naphthalene (99%) were obtained from Merck, India Ltd., S D Fine-Chem Limited, Mumbai and LobaChemie, India Ltd. respectively. Acetonitrile, dichloromethane and ethyl acetate (both HPLC grade and hypergrade/MS grade) were procured from Merck, India.

2.2 *Development of HPLC Methods for Identification and Quantification of Selected PAHs and Heterocyclics*

All the samples were run at a flow rate of 0.5 mL/min under isocratic mode with 80:20 (v/v) acetonitrile:water as the mobile phase. Detection of carbazole was carried out using a Photodiode Array Detector (PDA) at 230 nm wavelength. In the case of individual identification and quantification of fluorene and dibenzothiophene, 0.5 mL/min mobile phase flow rate was used under isocratic mode with 95:5 (v/v) acetonitrile:water as the mobile phase. Detection of both compounds was carried out using a Photodiode Array Detector (PDA) at 254 nm wavelength. HPLC method development for simultaneous identification and quantification of a mixture of nine compounds (containing three PAHs and six heterocyclics (nitrogen and sulphur) PAHs) was performed by modifying the method specified by Saha et al. (2010). The selected compounds were benzene, naphthalene, fluorene, pyridine, quinoline, carbazole, thiophene, benzothiophene and dibenzothiophene. The compounds were dissolved in acetonitrile and were run in the HPLC (410 Prostar Binary LC with 500 MS IT Photodiode Array Detectors, Varian Inc., USA) using this method. Individual compounds were also analyzed using the same method to determine their retention time. The developed method is summarized in Table 1. This method was subsequently used for simultaneous identification and quantification of various mixture combinations of fluorene, carbazole and dibenzothiophene in different experiments. Detection of all the compounds was carried out using a PDA at 254 nm wavelength.

Table 1 HPLC method for identification of mixtures of PAHs and heterocyclic PAHs

Time (min)	%A (Aqueous)	%B (Organic)	Flow (ml/min)
0	95	5	
195	5		
6	20	80	
9	20	80	
15	5	95	5
20	5	95	
21	95	5	
25	95	5	

2.3 *Development of GC-FID Methods for Identification and Quantification of Selected PAHs and Heterocyclics*

Mixed standards of three compounds namely, fluorene (PAH), carbazole and dibenzothiophene (heterocyclic PAHs) were prepared in the concentration range of 1–100 mg/L in dichloromethane. These mixed standards were run using a gas chromatograph equipped with a flame ionization detector (GC-FID) (Clarus 500, Perkin Elmer, USA). An Elite-5 (Perkin Elmer, USA) capillary column (length 30 m, internal diameter 0.53 mm and film thickness 1.5 μm) was used for running the standards in order to calibrate the instrument and to determine the limit of detection (LOD) and limit of quantitation (LOQ) of the individual compounds. Nitrogen was applied as the carrier gas and the injection volume was 1 μL. The instrument was run in a splitless mode as well as in split mode with a 10:1 split ratio. The injector and detector temperatures were 200 °C and 280 °C, respectively. The temperature program included an initial hold time of 2 min at an initial temperature of 140 °C followed by a ramp rate of 20 °C/min until 280 °C (3 min hold). The total run time was 12 min. The instrument was calibrated by running the mixed standards. LOD and LOQ for each of the compounds were determined by running the mixed standards in the 1–10 mg/L concentration range.

2.4 *Development of GC-GC-TOF-MS Methods for Identification and Quantification of Selected PAHs and Heterocyclics*

One-dimensional gas chromatography (1D GC) based methods were developed for identifying and quantifying benzene and PAHs (naphthalene and fluorene), PANHs (pyridine, quinoline and carbazole) and PASHs (thiophene, benzo[b]thiophene and dibenzothiophene) using a Leco Pegasus 4D GC-GC-TOF MS. Combination of Rxi 5-MS (30 m) primary column and Rxi 17Sil MS (2 m) secondary column (both

Restek, Made in USA) were used for developing the identification and quantification methods for these compounds. Calibration curves of the compounds were also prepared by running their standards between 1 and 10 mg/L concentration ranges. Dichloromethane was used as the solvent for dissolving all the compounds. A 1D GC method (for generating prominent compound peak) could be developed for all 9 compounds, however, calibration curves followed linearity only for 6 of the compounds namely, naphthalene, fluorene, quinoline, carbazole, benzothiophene and dibenzothiophene. After that, a mixture of all 6 compounds was analyzed using 1D GC and a method was developed for identifying each of them. The same method was applied for the low boiling point compounds, benzene, pyridine and thiophene, but none of them could be detected. Subsequently, a separate 1D GC method was developed for identifying pyridine, thiophene and benzene, individually and in ternary mixtures. Finally, a 1D GC method was developed for the identification of all the 9 compounds in a novenary mixture. Table 4 summarizes the details of all the 1D GC methods developed for the identification of all the compounds when applied individually and in mixtures.

3 Results and Discussion

3.1 Development of HPLC Methods for Identification and Quantification of Selected PAHs and Heterocyclics

Using the HPLC method for individual identification and quantification of fluorene, the retention time was 6.90 ± 0.09 min (mean ± SD) and the limit of detection (LOD) and limit of quantitation (LOQ) were 0.63 and 1.9 mg/L, respectively. ICH (International Committee on Harmonization) guidelines (ICH 2005) were followed for determining the LOD and LOQ values of the compounds. The extraction efficiency was 98.11 ± 1.00%. Using the HPLC method used for dibenzothiophene, the values of the parameters were: retention time 8.68 ± 0.04 min (mean ± SD), LOD 0.58 mg/L, LOQ 1.75 mg/L and extraction efficiency 99.12 ± 1.59%.

Retention times of each compound when they were run in a mixture were: benzene (0.5 min), pyridine (11.8 min), thiophene (13.5 min), quinoline (14.1 min), carbazole (15.6 min), benzothiophene (16.3 min), naphthalene (17.2 min), fluorene (19.9 min) and dibenzothiophene (20.8 min). The retention time of each compound did not vary much across the runs except for pyridine. Figure 1 represents a typical HPLC chromatogram of a mixture containing all nine compounds.

Table 2 summarizes the parameter values for individual substrates (carbazole, fluorene and dibenzothiophene) when the HPLC method developed for the mixture of the three compounds was performed.

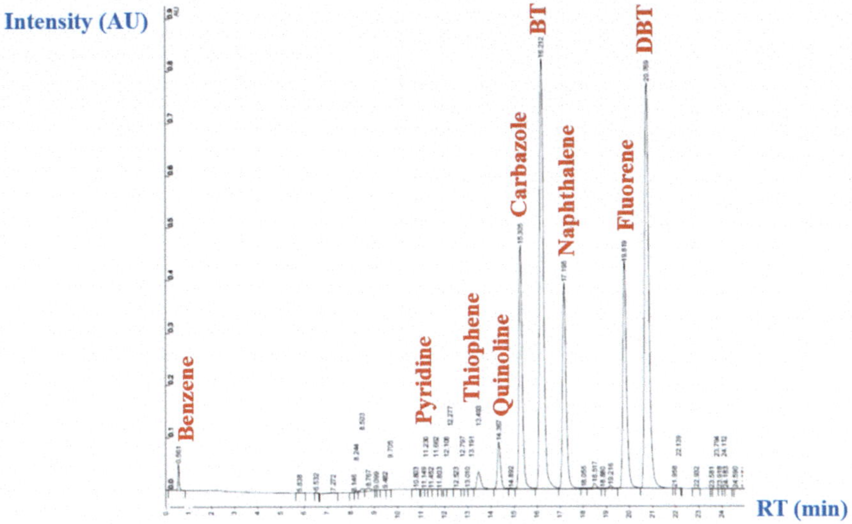

Fig. 1 HPLC chromatogram of a mixture of PAHs, PANHs and PASHs

Table 2 Summary of various parameter values for the HPLC method developed for mixtures of carbazole, fluorene and dibenzothiophene

Compounds	Retention time (min)	Extraction efficiency (%)	LOD (mg/L)	LOQ (mg/L)
Carbazole	15.67 ± 0.19	100 ± 5.55	1.19	3.61
Fluorene	20.23 ± 0.17	97.01 ± 1.02	0.67	2.02
Dibenzothiophene	21.27 ± 0.21	102.03 ± 0.99	1.39	4.20

3.2 Development of GC-FID Methods for Identification and Quantification of Selected PAHs and Heterocyclics

When a 10:1 split was provided, the linearity of the calibration curve for dibenzothiophene was poor, however, under splitless injection, calibration curves of all the compounds were proper within the 1–100 mg/L concentration range. A proper calibration curve could be developed for each of the compounds in the ternary mixture injected in splitless mode. LOD and LOQ for each of the compounds are listed in Table 3.

Table 3 Summary of LOD and LOQ of individual compound in mixtures

Compounds	LOD (mg/L)		LOQ (mg/L)	
	Mean	SD	Mean	SD
Carbazole	2.18	0.31	6.62	0.93
Fluorene	8.32	1.89	37.30	11.33
Dibenzothiophene	9.71	3.67	29.43	11.33

Table 4 Summary of 1D GC methods for identification of all PAHs and heterocyclic PAHs

Class	Compound/ mixture of compounds	Td (sec)	Tim (°Q)	thl (min)	RRi (°C/ min)	Tint (°C)	th2 (min)	RR: (°C/ min)	Tfin (°C)	th3 (min)	Mode
	Benzene (B)	150	40	10				1.5	90	1	1:10 split
PAHs	Naphthalene (N)	350	50	10				30	250	5	Splitless
	Fluorene (F)	350	70	3				30	280	5	Splitless
	Pyridine (P)	180	40	3				5	100	0.5	Splitless
PANHs	Quinoline (Q)	200	70	3				30	280	5	Splitless
	Carbazole (CBZ)	350	70	3				40	280	5	Splitless
	Thiophene (Th)	150	40	10				1.5	90	1	1:10 split
PASHs	Benzo[b]thiophene (BT)	350	50	10				30	250	5	Splitless
	Dibenzothiopliene (DBT)	350	70	3				30	280	5	Splitless
	N + F + Q T CBZ – BT – DBT	200	50	5				10	200	5	Splitless
	B + P – Th	150	40	10				1.5	90	1	Splitless
	PAHs + PANHs + PASHs	150	40	10	1.5	90	3	10	280	1	Splitless

Note t_d: Acquisition delay time, T_{ini}: Initial oven temperature, t_{h1}: Initial hold time, RR_1: First intermediate temperature ramp rate, T_{int}: First intermediate oven temperature, t_{h2}: First intermediate hold time, RR_2: Final temperature ramp rate, T_{fin}: Final oven temperature and t_{h3}: Final hold time

3.3 Development of GC-GC-TOF MS Methods for Identification and Quantification of Selected PAHs and Heterocyclics

Following the 1D GC methods for the individual compounds as described in Table 4, the retention time of the individual compounds along with their boiling point are listed in Table 5.

Table 5 Retention times and boiling point of the compounds in the GC column

Compound name	Retention time (s)	Boiling point (°C)
Carbazole	2388	359
Quinoline	1306	237
Pyridine	254	115
Fluorene	1971.5	295
Naphthalene	1193	218
Benzene	167.98	80
Dibenzothiophene	2254	332
Benzotliiophene	1209	221
Thiophene	173.76	84

It can be seen from Table 5 that the boiling points of pyridine, benzene and thiophene are much lower compared to the other compounds. The solvent, dichloromethane, has a boiling point of 39.6 °C, which is closer to the boiling points of these 3 compounds (which is not the case with the remaining compounds). Hence, a slower ramp rate (5 °C/min) was required for the identification and quantification of pyridine. Again, the boiling points of benzene and thiophene are very close and are less than that of pyridine. Hence, the ramp rate had to be further reduced (1.5 °C/min) to resolve these components. Using the method for identification of a mixture of the compounds having low boiling points, the retention times (sec) were benzene (172.1), pyridine (269.592) and thiophene (177.58). Using the method for identification of all the compounds in a mixture, the retention times (sec) were: benzene (172.52), naphthalene (2388.12), fluorene (3344.88), pyridine (269.592), quinoline (2686.16), carbazole (3534.62), thiophene (177.52), benzothiophene (2430.35) and dibenzothiophene (3478.42). The total ion chromatogram pertaining to the mixture containing all 9 compounds is shown in Fig. 2. Compounds with lower boiling points eluted very early as can be seen from the figure. The elution sequence of the compounds was as per their boiling points. The peaks were close for those compounds having closer boiling points (such as benzene and thiophene and naphthalene and benzothiophene). Hence, knowing the boiling points of solvents and compounds of interest is important for the development of a GC method. For compounds having boiling points close to that of the solvent, higher holding time at lower initial temperature and slower ramp rate could effectively separate the peaks.

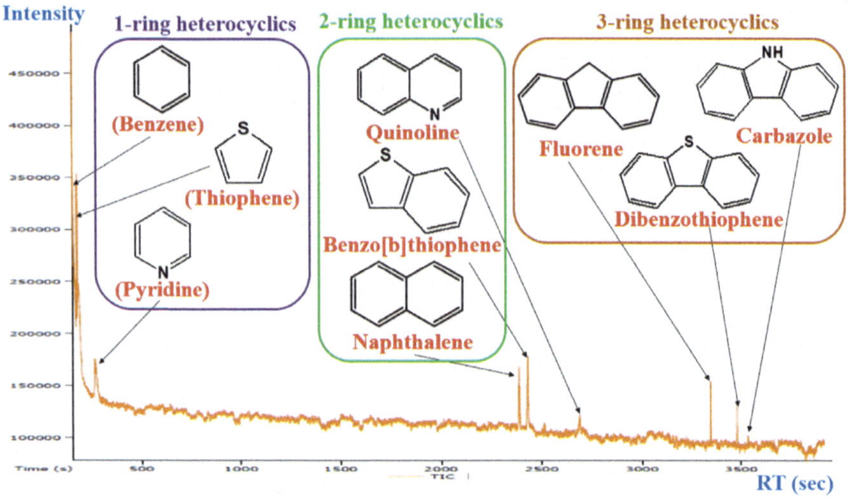

Fig. 2 Total Ion Chromatogram (TIC) of a mixture of 9 PAHs and heterocyclic PAHs and their individual peaks analysed using a GC-GC-TOF MS operated in 1D mode

4 Conclusion

The mobile phase program developed for HPLC analysis could clearly separate the peaks of the PAHs and heterocyclics in the mixtures. The peaks were also prominent and sharp. The method was also suitable for identifying the compounds when applied at low concentrations (about 2–10 mg/L). As observed from the LOD and LOQ values of the compounds using the GC-FID method, it can be concluded that the method is suitable for the identification and quantification of the compounds when applied at 10–100 mg/L. From the 1D GC method development for simultaneous identification and quantification of the PAH and heterocyclic PAH mixtures, it can be concluded that boiling points of the compounds relative to the solvent (used for dissolving them) are very important. Proper holding time and ramping rate in the GC oven temperature program help to resolve the peaks of compounds having similar boiling points. For compounds having boiling points close to the solvent, a higher initial holding time at a lower initial oven temperature and a slower ramp rate can effectively separate their peaks.

Acknowledgements The authors acknowledge SAIF, IIT Bombay, for providing the facilities for GC-GC-TOF MS and HPLC analyses.

References

Aislabie, J., Bej, A.K., Hurst, H., Rothenburger, S. and R.M. Atlas (1990). Microbial degradation of quinoline and methylquinolines. *Appl. Environ. Microbiol.*, **56**: 345–351.

Akdogan, H.A. and N.K. Pazarlioglu (2011). Fluorene biodegradation by *P. ostreatus*—Part I: Biodegradation by free cells. *Process Biochem.*, **46**: 834–839.

Bahrami, A., Shojaosadati, S.A. and G. Mohebali (2001). Biodegradation of dibenzothiophene by thermophilic bacteria. *Biotechnol. Lett.*, **23**: 899–901.

Bai, Y., Sun, Q., Xing, R., Wen, D. and X. Tang (2010). Removal of pyridine and quinoline by bio-zeolite composed of mixed degrading bacteria and modified zeolite. *J. Hazard. Mater.*, **181**: 916–922.

Berthou, F. and V. Vignier (1986). Analysis and fate of dibenzothiophene derivatives in the marine environment. *Int. J. Environ. Anal. Chem.*, **27**: 81–96.

Bezalel, L., Hadar, Y., Fu, P.P., Freeman, J.P. and C.E. Cerniglia (1996). Initial oxidation products in the metabolism of pyrene, anthracene, fluorene, and dibenzothiophene by the white rot fungus *Pleurotus ostreatus*. *Appl. Environ. Microbiol.*, **62**: 2554–2559.

Bleeker, E.A.J., Wiegman, S., de Voogt, P., Kraak, M., Leslie, H.A., de Haas, E. and W. Admiraal (2002). Toxicity of azaarenes. *Rev Env. Contam. Toxicol.*, **173**: 39–83.

Byrd, G.D., Fowler, K.W., Hicks, R.D., Lovette, M.E. and M.F. Borgerding (1990). Isotope dilution gas chromatography-mass spectrometry in the determination of benzene, toluene, styrene and acrylonitrile in mainstream cigarette smoke. *J. Chromatogr. A*, **503**: 359–368.

Celis, R., Real, M., Hermosín, M.C. and J. Cornejo (2006). Desorption, persistence, and leaching of dibenzofuran in European soils. *Soil Sci. Soc. Am. J.*, **70**: 1310–1317.

Chupungars, K., Rerngsamran, P. and S. Thaniyavarn (2009). Polycyclic aromatic hydrocarbons degradation by *Agrocybe* sp. CU-43 and its fluorene transformation. *Int. Biodeterior. Biodegrad.*, **63**: 93–99.

Collina, E., Bestetti, G., Di Gennaro, P., Franzetti, A., Gugliersi, F., Lasagni, M. and D. Pitea (2005). Naphthalene biodegradation kinetics in an aerobic slurry-phase bioreactor. *Environ. Int.*, **31**: 167–171.

Eaton, R.W. and J.D. Nitterauer (1994). Biotransformation of benzothiophene by isopropylbenzene-degrading bacteria. *J. Bact.*, **176(13)**: 3992–4002.

Flego, C. and C. Zannoni (2011). N-containing species in crude oil fractions: An identification and quantification method by comprehensive two-dimensional gas chromatography coupled with quadrupole mass spectrometry. *Fuel*, **90**: 2863–2869.

Gai, Z., Yu, B., Wang, X., Deng, Z. and P. Xu (2008). Microbial transformation of benzothiophenes, with carbazole as the auxiliary substrate, by *Sphingomonas* sp. strain XLDN2-5. *Microbiology*, **154**: 3804–3812.

Geier, M.C., Chlebowski, A.C., Truong, L., Simonich, S.L.M., Anderson, K.A. and R.L. Tanguay (2018). Comparative developmental toxicity of a comprehensive suite of polycyclic aromatic hydrocarbons. *Arch. Toxicol.*, **92**: 571–586.

Ghosh, P. and S. Mukherji (2020). Degradation of carbazole, fluorene, dibenzothiophene and their mixture by *P. aeruginosa* RS1 in petroleum refinery wastewater. *J. Water Process Eng.*, **37**: 10154.

Gieg, L.M., Otter, A. and P.M. Fedorak (1996). Carbazole degradation by metabolic characteristics and the identification of some metabolites. *Environ. Sci. Technol.*, **30(2)**: 575–585.

Gilbert, S.C., Morton, J., Buchanan, S., Oldfield, C. and A. McRoberts (1998). Isolation of a unique benzothiophene-desulphurizing bacterium, *Gordona* sp. strain 213E (NCIMB 40816), and characterization of the desulphurization pathway. *Microbiology*, **144**: 2545–2553.

Grifoll, M., Selifonov, S.A. and P.J. Chapman (1995). Transformation of substituted fluorenes and fluorene analogs by *Pseudomonas* sp. strain F274. *Appl. Environ. Microbiol.*, **61**: 3490–3493.

Hammer, E., Krowas, D., Schäfer, A., Specht, M., Francke, W. and F. Schauer (1998). Isolation and characterization of a dibenzofuran-degrading yeast: Identification of oxidation and ring cleavage products. *Appl. Environ. Microbiol.*, **64**: 2215–2219.

Haritash, A.K. and C.P. Kaushik (2009). Biodegradation aspects of polycyclic aromatic hydrocarbons (PAHs): A review. *J. Hazard. Mater.*, **169**: 1–15.

Huang, H., Bowler, B.F.J., Zhang, Z., Oldenburg, T.B.P. and S.R. Larter (2003). Influence of biodegradation on carbazole and benzocarbazole distributions in oil columns from the Liaohe basin, NE China. *Org. Geochem.*, **34**: 951–969.

International Committee on Harmonization. Validation of analytical procedures: text and methodology. Q2 (R1), 2005. 05.

Jensen, A.-M., Finster, K.W., Karlson, U., 2003. Degradation of carbazole, dibenzothiophene, and dibenzofuran at low temperature by *Pseudomonas* sp. strain C3211. Environ. Toxicol. Chem. 22, 730–735.

Jianlong, W., Xiangchun, Q., Liping, H., Yi, Q., Hegemann, W., 2002. Microbial degradation of quinoline by immobilized cells of *Burkholderia pickettii.* Water Res. 36, 2288–2296.

Keith, L.H., 2015. The source of US EPA's sixteen PAH priority pollutant. Polycycl. Aromat. Compd. 35, 147–160.

Lee, S.T., Rhee, S.K. and G.M. Lee (1994). Biodegradation of pyridine by freely suspended and immobilized *Pimelobacter* sp. *Appl. Microbiol. Biotechnol.* **41**: 652–657.

Lin, C., Gan, L. and Z.L. Chen (2010). Biodegradation of naphthalene by strain *Bacillus fusiformis* (BFN). *J. Hazard. Mater.*, **182**: 771–777.

Meyer, S. and H. Steinhart (2000). Effects of heterocyclic PAHs (N, S, O) on the biodegradation of typical tar oil PAHs in a soil/compost mixture. *Chemosphere*, **40**: 359–367.

Mollea, C., Bosco, F. and B. Ruggeri (2005). Fungal biodegradation of naphthalene: Microcosms studies. *Chemosphere*, **60**: 636–643.

Mosmeri, H., Alaie, E., Shavandi, M., Dastgheib, S.M.M. and S. Tasharrofi (2017). Bioremediation of benzene from groundwater by calcium peroxide (CaO_2) nanoparticles encapsulated in sodium alginate. *J. Taiwan Inst. Chem. Eng.*, **78**: 299–306.

Mundt, M. and J. Hollender (2005). Simultaneous determination of NSO-heterocycles, homocycles and their metabolites in groundwater of tar oil contaminated sites using LC with diode array UV and fluorescence detection. *J. Chromatogr. A*, **1065**: 211–218.

Rajagopal, S., Grimm, T.L., Collins, D.J. and R. Miranda (1990). A practical scheme for the GC analysis of pyridine hydrodenitrogenation reaction products. *Anal. Lett.*, **23**: 649–57.

Rivas, I.M., Mosbæk, H. and E. Arvin (2003). Product formation from thiophene by a mixed bacterial culture. Influence of benzene as growth substrate. *Water Res.*, **37**: 3047–3053.

Saha, S., Mistri, R. and B.C. Ray (2010). Determination of pyridine, 2-picoline, 4-picoline and quinoline from mainstream cigarette smoke by solid-phase extraction liquid chromatography/ electrospray ionization tandem mass spectrometry. *J. Chromatogr. A*, **1217**: 307–311.

Salam, L.B., Ilori, M.O. and O.O. Amund (2017). Properties, environmental fate and biodegradation of carbazole. *3 Biotech*, **7**: 1–14.

Stumpf, Á., Tolvaj, K. and M. Juhász (1998). Detailed analysis of sulfur compounds in gasoline range petroleum products with high-resolution gas chromatography-atomic emission detection using group-selective chemical treatment. *J. Chromatogr. A*, **819**: 67–74.

Švábenský, R., Kočí, K. and Z. Šimek (2007). The study of properties of HPLC determination of polycyclic aromatic nitrogen heterocycles. *Int. J. Environ. Anal. Chem.*, **87**: 337–349.

Traulsen, F., Andersson, J.T. and M.G. Ehrhardt (1999). Acidic and non-acidic products from the photo-oxidation of the crude oil component dibenzothiophene dissolved in seawater. *Anal. Chim. Acta*, **392**: 19–28.

van Afferden, M., Schacht, S., Klein, J. and G.T. Hans (1989). Degradation of dibenzothiophene by *Brevibacterium* sp.DO. *Arch. Microbiol.*, **153**: 324–328.

Wang, X., Gai, Z., Yu, B., Feng, J., Xu, C., Yuan, Y., Lin, Z. and P. Xu (2007). Degradation of carbazole by microbial cells immobilized in magnetic gellan gum gel beads. *Appl. Environ. Microbiol.*, **73**: 6421–6428.

Chapter 9
A Simple Approach for Designing a Waste Collection System for Guwahati City Using Google Earth

Abhishek Singhal and Brajesh Kumar Dubey

1 Introduction

Economic transformations in the tropical Asian region since the mid-twentieth century have greatly affected the rate of manufacture and consumption of various resources (MacRae 2012). This massive increment in resource consumption leads to huge quantities of MSW which is quite a problem in most of the countries in the region (Agamuthu et al. 2009; Khan and Samadder 2016). In most developing countries, including India, there is a lack of an organized door-to-door collection of waste which has led to littering and illegal dumping habits. Local municipalities are the main service providers for municipal solid waste management (MSWM) and waste collection and transport (WC&T) is an integrated part of it. However, due to a lack of infrastructure and scientific approach, many of the local authorities in developing and underdeveloped nations are failing to deal with their massive amount of MSW (Annepu 2012; Goel 2008; Sharholy et al. 2008). In many developing and underdeveloped countries, WC&T comprises approximately 80–90% of the overall MSWM budgets while waste collection efficiencies remain less than 50% (Hoonberg and Bhada-Tada 2012; Dutta and Goel 2017; Ghose et al. 2006). Due to huge expenditures on waste collection, lack of financial resources and poor waste collection efficiency, illegal dumping/littering, and burning of garbage are among the most prominent problems in urban India.

As the WC&T system often accounts for the bulk of the SWM budgets, even a tiny portion of the improvement in the waste collection operation may lead to

A. Singhal (✉)
Department of Material Science and Environmental Engineering, Tampere University, Tampere, Finland
e-mail: abhisheksinghal.iitkgp@gmail.com

A. Singhal · B. K. Dubey
Department of Civil Engineering, Indian Institute of Technology Kharagpur, Kharagpur 721 302, India

S. Goel (ed.), *Advances in Solid and Hazardous Waste Management*,
https://doi.org/10.1007/978-3-031-49144-3_9

substantial cost savings (Tavares et al. 2009; Xue et al. 2015). Therefore, optimization of the WC&T is a critical need in developing countries for economically and environmentally sustainable MSWM system planning. In past, several studies were conducted on optimization of waste collection and vehicle routing by using different algorithms, modelling/models, mathematical approaches/problems, and Remote Sensing-Geographical Information system (RS-GIS) based approaches. A summary of methodologies and tools used in the above-mentioned studies are summarized in Table 1. Several researchers have applied different RS and GIS analyses for collection routes optimization out of which network analyst extension of Esri's ArcGIS is utilized more often.

For developing a WC&T plan or for collection route optimization, some basic amount of data is essential such as road network and road quality data, historical traffic data, elevation data, bin and truck size availability, etc. However, even in the large cities of many developing or underdeveloped nations, these basic data are still

Table 1 Various approaches used for optimizing MSW collection and transportation

Authors	Work done and methodology used
Lin et al. (2011)	MSW collection done by using Ant colony optimization (ACO) algorithm to provide flexibility to local residents
Lin and Kao (2008)	MSW collection done by using mixed-integer optimization model considering factors such as road network integrity, collection cost, compactness, and regional proximity
O'Connor (2013)	MSW Collection Vehicle Route Optimization using ArcGIS-Network Analyst and Vehicle routing problem solver
Malakahmad et al. (2014)	MSW collection routes design and their optimization by Territorial analysis and ArcGIS-Network Analyst
Das and Bhattacharyya (2015)	Optimization of MSW collection routes with travelling salesman problem (TSP)
Vecchi et al. (2016)	Mathematical modelling and Linear programming for optimizing MSW collection routes
Laureri et al. (2016)	A heuristic procedure was developed for the optimal planning of wet waste collection
Khan and Samadder (2016)	Allocation of MSW collection bins and route optimisation using ArcGIS (Network Service area solver and Network Analyst)
Lella et al. (2017)	MSW collection and transportation optimization using vegetation land cover estimation (NDVI mapping) and ArcGIS-Network Analyst
Hannan et al. (2018)	MSW collection and route optimization using a modified particle swarm optimization (PSO) algorithm
Vu et al. (2019)	Combining of artificial neural network (ANN) waste prediction model with GIS route optimization for MSW collection
Singhal and Goel (2021a)	Developed an integrated solid waste management (ISWM) plan using Google Earth and optimized using linear programming tools

not available or incomplete, or very difficult to recover from the local authorities. Many big cities in India still lack good quality historical traffic data which is vital for selecting waste collection timings for the commercial and traffic-congested areas. Also, for developing a WC&T system and for monitoring, considerable knowledge regarding GIS software and traffic and vehicle fleet management is crucial. For utilizing mathematical and linear programming models as shown in Table 1, a great deal of mathematical and computer programming skills is required for their implementation. But gaining such knowledge and handling GIS software such as Arc-GIS and QGIS, requires skilled people. Handling such complex software can be highly challenging for people responsible for waste collection in the region, especially for small to moderate cities and towns in developing nations such as India. Furthermore, for efficiently running these GIS software and network analysis programs, computer systems with high performance and high internet speed are crucial. This may easily become a financial burden to the local authorities which are already lacking financial resources. So, an easy approach that requires the least amount of data and less sophisticated GIS software for developing and monitoring the WC&T system, can greatly help the local municipalities in efficiently collecting the waste. Google Earth and Google Maps can be an easy and possible solution to this problem.

Google Earth is a computer program that renders a 3D representation of the Earth which is completely based on remotely sensed satellite imageries (Hu et al. 2013; Yu and Gong 2012). Nowadays researchers have been using Google Earth to visualize data for studying various environmental issues and phenomena like time series analysis of dumpsites (Singhal and Goel 2020, 2021b), air pollution modelling (Liu and Kenjeres 2017), environmental impact analysis (Singhal et al. 2020; Singhal and Goel 2021c) and local weather patterns study (Kumar and Mutanga 2018), etc. In Google Earth, available images can be zoomed in to explore major urban areas where a resolution of about 1 m/pixel is available. Several handful tools are available in Google Earth such as mapping and labelling of different locations, area and distance calculation tools, historical images, etc. Using the abovementioned and many other sophisticated tools in Google Earth, it can be used for environmental studies and generating data, especially where detailed RS, GIS and local data is scarce. Google Earth can easily run at basic internet speed (1–2 Mbps) and perform efficiently even on basic computer systems. Even though, compared to present optimization and vehicle routing options, manual routing is relatively less efficient (Worrell and Vesilind 2012). However, using together with Google and common sense judgement, the efficiency of the manual routing can be increased manifold. Singhal and Goel (2021a) have used Google Earth with linear programming to optimize the waste collection routes for a small study area. Besides Google Earth, Google maps can also be used for designing the WC&T system due to its historical and live traffic data of the road network. Using this data, heavy traffic areas can be avoided at peak traffic hours and less traffic-congested roads can be selected for routing. In the available body of literature, only a few studies have utilized Google Earth for waste collection and management. Furthermore, no study was found that utilized Google Maps for waste collection routing and fleet management. Due to being freely available on the internet, both Google Earth and Google Maps have a high potential to be utilized for developing a low-cost but efficient WC&T system.

The present study aimed to develop and illustrate a simple methodology for developing a WC&T system for an urban area by using a minimum amount of external data and easily accessible RS-GIS software. Division-IV of Guwahati City was selected as the study area in the present study. Google Earth was used in the present study for designing bin location plans and sketching waste collection routes. First per capita waste generation, population and waste composition data of the study area were collected. Based on these data bin sizes for the segregated bin were calculated and service areas for bins were designed. Based on the calculated service area, bins were located in the study area and a bin location plan was designed for secondary waste collection by using Google Earth Pro. After designing the bin location plan, routes were created manually for the vehicle routing and the time required for the waste collection was calculated. Finally, using historical traffic data from Google Maps, time duration and routes for each vehicle were assigned for waste collection. The approach used in the study is very simple, and the software used is freely available and easy to operate. This methodology can be replicated and extended to cover an entire city or cities for designing an efficient waste collection system.

1.1 Description of Study Area and Current Waste Collection Practices

Guwahati city also known as the City of Eastern Light, is the capital city of Assam state as well as the biggest city in the state with a population of 957,352 (census 2011). Guwahati city is situated on the southern banks of the Brahmaputra River (on 26°10′20″ North and 91°44′45″ East geographical coordinates). The general topographic features of the city are hilly and lie between the ground elevations ranging from about 40 m near Brahmaputra to 575 m towards Meghalaya hills constituting the southern boundary of the city. The city receives an annual rainfall of 152–324 cm, with fairly heavy rains from May to October. The humidity prevails throughout the year but is usually highest during July and August (76–94%).

Guwahati Municipal Corporation (GMC) is responsible for solid waste management in the city. The entire GMC area is divided into 6 divisions from where about 800 tonnes of MSW are generated on a daily basis (Fig. 1a). Collection of MSW is done by GMC as a primary and secondary collection of the MSW. The primary collection consists of door-to-door collection of the MSW with source segregation into two streams: wet (biodegradable and inert waste) and dry waste (recyclables). The result of the other study conducted parallel with the present study shows that around 45% of the waste generated in the city is organic waste and 55% of the waste is recyclables and inert waste material (Fig. 3). Detailed waste characterization results are included in the study by Singhal et al. (2021a). After primary collection waste is transferred to the nearest community bin with the help of tricycles and small tipper trucks. During secondary waste collection, waste from the community bins, storage depots and transfer stations is collected by using garbage compactor trucks. Even though proper source segregation is good practice in many areas during primary collection but at the end of the day, all the waste is mixed together and put into a single community bin. Currently, there is no waste treatment and sorting facility in the city due to which all of the waste is crudely dumped at the Boragaon dumpsite. Presently, the systems applied by the municipal authorities for waste collection are inefficient and unscientific. Furthermore, existing waste collection bins are located sparsely, which are not easily accessible and are mostly overfilled due to inadequate collection frequency. Present MSWM practices in the region are unsustainable and unhygienic which needs immediate attention.

The focus of the study was division IV as it includes the main market area of the city, many residential areas/colonies and both hilly and plain terrain (Fig. 1b). The total population of division-IV is 148,336 which generates about 82 tonnes of waste on daily basis. It was selected for the study because it was the most traffic-congested area in the entire city and due to poor waste collection requires the most attention. Also, as this division is a good representative of a busy metropolitan area of a typical Indian city, the approach used in division-IV can be extended to cover the entire Guwahati city or any other metropolitan areas.

2 Materials and Methodology

Flow chart of the development of the waste collection system is shown in Fig. 2.

2.1 Data Collection

The first step in designing of waste collection system was the calculation of the bin size required for the storage of waste. Waste composition, per capita waste generation, and population data are vital for calculating bin size. So, from GMC ward wise population data, division wise map and per capita waste generation data were collected. Waste composition data was used from MSW characterization studies conducted parallel with the present study (Singhal et al. 2021a). The loading time of the bin while waste collection and unloading time of the truck at the transfer station were noted during the site survey. These parameters are crucial for designing vehicle routes for efficient waste collection.

2.2 *Bin Size Calculation*

Choosing an appropriate bin size avoids overfilling and illegal dumping of waste. For calculating bin sizes certain assumptions were made on the basis of literature, site survey and according to the city demands. All assumptions are noted here:

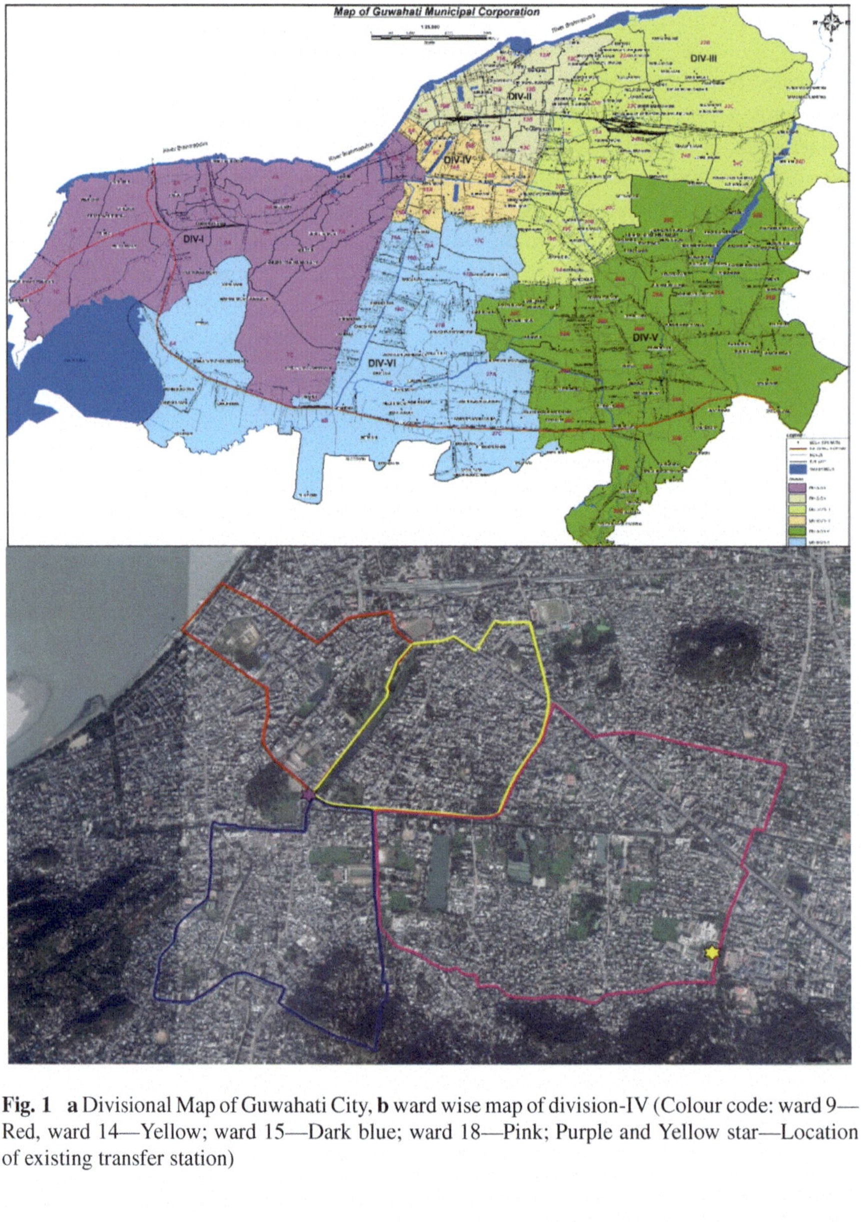

Fig. 1 **a** Divisional Map of Guwahati City, **b** ward wise map of division-IV (Colour code: ward 9—Red, ward 14—Yellow; ward 15—Dark blue; ward 18—Pink; Purple and Yellow star—Location of existing transfer station)

1. Due to door-to-door collection in the primary collection of generated MSW, large bins are mainly required to transfer the waste collected during the primary collection in the nearby area. So, the bin plan was mainly designed for secondary waste collection.
2. Currently 3 m^3 bins without source segregation are used in the city which serves approximately 1230 people. However, at each location, 2 separate bins are provided for source segregation. One bin is a "Dry waste bin" for the storage of recyclables and the other bin is a "Wet waste bin" for the storage of organic household waste.
3. The size of the bin depends on the density of MSW and the amount of waste generated along with a factor of safety. The density of the wet waste generated in the city is ranging from 500 to 700 kg/m^3 and for dry waste density is ranging from 300 to 450 kg/m^3. As per Goel (2008), for designing an efficient waste storage system, a per capita generation rate at the higher end and a low bulk density should be assumed (Goel 2008). So, for safety and to avoid overfilling even in extreme cases, 500 kg/m^3 density for wet waste and 300 kg/m^3 density for dry waste was selected for design purposes.
4. Current per capita solid waste generation rate in the city is 0.5 kg per capita per day. To avoid overfilling of bins and for the waste storage system which has to work efficiently for the next 10–15 years, the per capita waste generation rate was selected slightly higher than the present i.e. 0.6 kg/capita/day. According to the Census of India 2011, the number of people per house was taken at 4.4.
5. A factor of safety (FOS) of 1.5 (50% extra space) is provided in every bin to avoid overloading even at the peak seasons (like festivals) and for providing extra space for the next 10 years in which per capita waste generation rates will also increase.

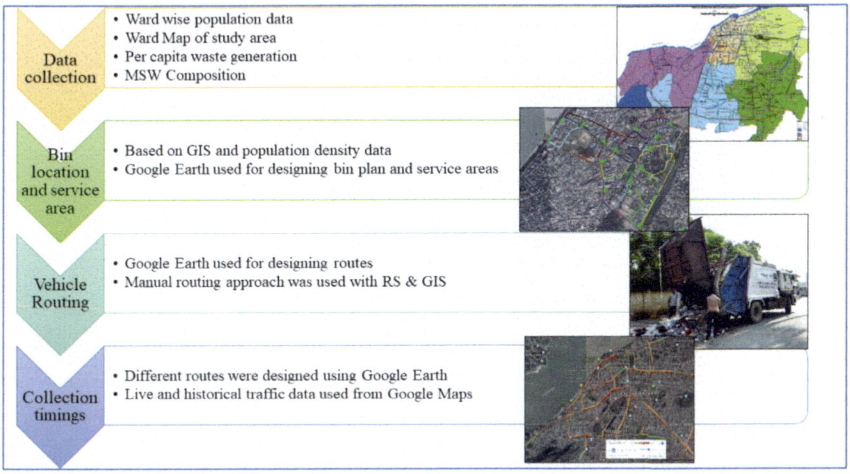

Fig. 2 Step wise flow chart for developing bin location plan and waste collection system

6. As the annual average temperature (11–37 °C) and moisture (57–83%) in the region is higher, a daily collection of MSW is very important for the hygienic environment. So, a daily collection of waste was implemented, and bin size was calculated according to that.
7. As per the results of the physical waste characterization shown in/collected from, the composition of the waste is assumed to be 45% wet waste (mainly organics) and 55% dry waste (recyclable and inert waste) by weight.

By using the above assumptions bin sizes for each wet and dry waste bin were calculated for the storage of waste for secondary collection. The formula used for bin size calculation was:

$$\text{Bin Size} = \textit{Quantity of the waste} \times \textit{Factor of safety}/\textit{Density of the waste} \quad (1)$$

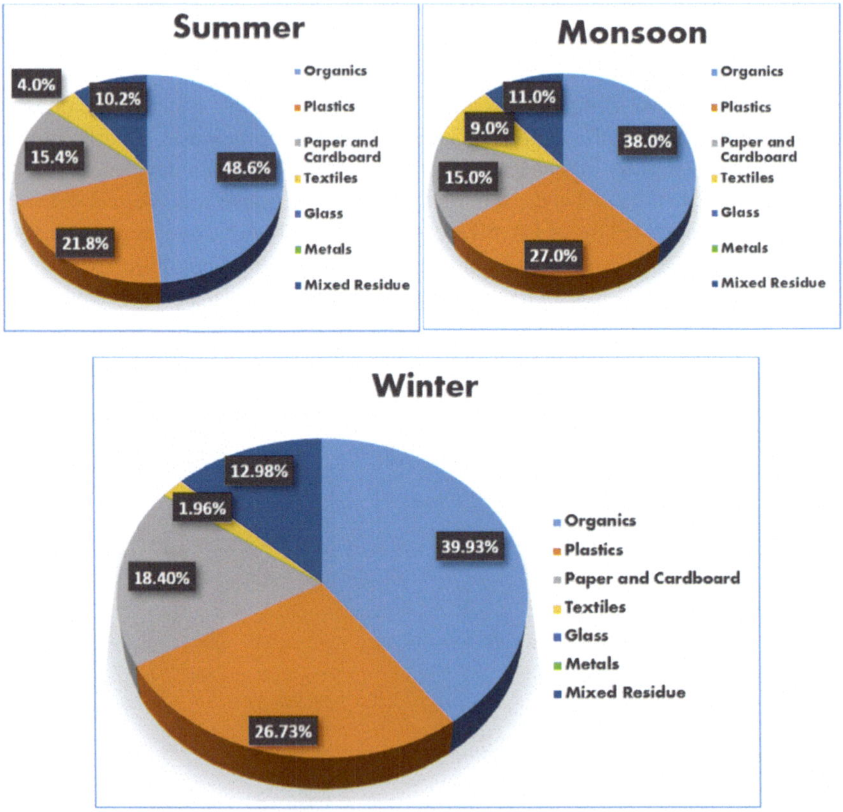

Fig. 3 Waste composition data of Guwahati City

2.3 Design of Bin Location Plan

After calculating bin size, the total number of bins required in division IV was calculated ward wise. A total number of bins required in a ward was calculated by dividing the total population of the ward by the population served by a wet waste and a dry waste bin. The service area of a single bin location (which represents a wet waste and one dry waste bin) was calculated using ward-wise population density data and the effective area of the ward. The effective area of a ward was calculated using the ruler toolbar in Google Earth Pro software by eliminating unhabituated spaces such as playgrounds, barren or empty landmass, water bodies, etc. from the total area of the ward. From the population data and effective ward area, the population density of the ward was calculated. A single bin location serves a total of 1230 people, so with help of population density data, an area in which 1230 people are living was calculated which is the serving area of a single bin location.

After calculating the service area of a bin location, each ward of division-IV was divided into the service areas manually using Google Earth Pro. To avoid complexity and provide flexibility in the primary collection, the service area of 2 bin locations was merged. For example, in ward-9 service area of a bin is 24000 m^2 but providing such a small area for the primary collection may restrict the primary collection vehicles (like tippers and tricycles-handcarts) which later will increase collection time and fuel consumption. So, by merging the service areas of two bin locations such conditions can be avoided and placing bins close to a major road will become easier. This will improve the secondary collection efficiency while ensuring efficient primary collection.

After sketching bin service areas, the location of bins was assigned to the ward map with the help of Google Earth Pro software. To reduce the collection time and for increasing collection efficiency, bins are located along the roadside, near road intersections and clubbed together wherever possible. Except few (~8%), most of the bins were placed alongside roads wider than two lanes where a garbage compactor truck can move easily. Also, 2 bin points (4 bins- 2 wet and 2 dry waste bins) are tried to provide together if the area is available alongside the road. This will reduce the number of stoppages which significantly reduces the collection time and effort.

2.4 Vehicle Routing for Secondary Waste Collection

In general, manual routing methods may result in inefficient utilization of resources and higher collection costs (when compared to other methods) when applied to large areas. However, by using tools/software such as Arc-GIS (Network-Analyst), and Google Earth and by following specific guidelines, the efficiency of manual routing can be increased manifolds. In the present study manual vehicle routing approach was followed by using Google Earth Pro software. Specific guidelines for manual vehicle routing for the waste collection were followed by Worrell and Vesilind (2012). Also, for drawing routes for secondary waste collection for division IV, certain assumptions and points were followed:

1. Route will start from the GMC garage and after completing the first route trucks will go to the transfer station. Then next route will begin from the transfer station itself and end at the same/another transfer station. At the end of the working day route will end at the GMC.
2. Collection system is assumed to be semi-automated with a crew size of two and waste collection frequency is on an everyday basis.
3. Two existing transfer stations will be used in the study for the storage of MSW collected from the study area. At the end of the day or on alternating days, waste can be transferred from the TS to the required locations (landfill, Material recovery facility-MRF, waste treatment facility) into a larger truck for transportation cost savings (Locations of TS shown in Fig. 5).
4. Garbage compactor truck selected is of size 14 m^3 and has a compaction ratio of 2.5:1 with a maximum density of 750 kg/m^3.
5. Average velocity of the truck was assumed to be 20 km/h, the unloading time of the truck at the transfer station is 10 min and the loading time of the bin is 5 min (vehicle speed, loading and unloading time were calculated during the survey). Also, 50% extra travelling time is provided during collection besides a 30 min break and two 15 min breaks. Extra time was also provided for unknown traffic conditions which can cause a significant delay during waste collection. This extra time can be reduced and further optimized after testing calculated routes for a month.
6. Waste collection and vehicle routing were avoided in heavy traffic areas during peak hours. These areas were covered in the early morning or during non-peak hours.

Using the above instructions and assumptions waste collection routes were designed using the path toolbar in Google Earth Pro software. In the present study, waste collection routes were only created for wet waste collection. Collection routes for dry waste collection can easily be drawn using the same methodology. Collection routes were designed such that all the bin locations in division-IV will be covered.

After designing waste collection routes total distance and total time taken by the waste collection trucks were calculated. Besides these parameters, another crucial parameter is required for selecting a time period for waste collection to avoid heavy traffic conditions. For evaluating traffic conditions in the study area, Google Maps was used to evaluate the traffic volume on the roads during different times of the day. Google Maps provides real-time as well as historical data of a particular road at specific times on specific days. A historical data map was used in the present study to assess the average traffic conditions of the roads in the study area. Except for Sunday on which traffic is relatively less in the daytime, no significant difference in the traffic volume was found in the rest of the days. So, routes were designed irrespective of days and they can be followed throughout the week. After collecting historic traffic volume data and calculating the required collection time, a time period for waste collection in a single day was assigned.

3 Results and Discussion

The main aim of the study was to provide a methodology for developing an efficient waste storage and collection system using easily accessible data and mapping software. The strategy consists of designing a bin location plan and solid waste collection system using Google Earth software and a manual vehicle routing approach.

3.1 Bin Location Plan

The results of the bin location plan show how to incorporate waste composition data, waste generation data and population data with Google Earth to design a solid waste storage system. For designing bin location plans first bin sizes required for the proper secondary MSW storage were calculated. Detailed calculations used for selecting bin size are shown below:

A single bin location containing 2 bins (wet waste and dry waste bin) are provided to serve 1230 people in the nearby area.

Total waste generated by the 1230 people = 1230 × 0.6 kg/capita/day = 738 kg.

Size required for wet waste bin: The composition of wet waste in the total waste is 45% by weight.

- Amount of wet waste = 738 kg × 0.45 = 332.1 kg
- Density of wet waste is 500 kg/m^3; Bin size required = 332.1 kg × 1.5 (FOS)/500 kg/m^3
- 0.996 m^3 → 1 m^3; *So, bin size required for the wet waste bin is 1* m^3.

Size required for dry waste bin: Composition of dry waste in the total waste is 55% by weight.

- Amount of dry waste = 738 × 0.55 = 405.9 kg
- Density of dry waste is 300 kg/m^3; Bin size required = 405.9 kg × 1.5 (FOS) / 300 kg/m^3
- 2.03 m^3 → 2 m^3; *So, bin size required for the wet waste bin is* 2 m^3.

So, at a single location, 2 bins are required of **size 1 m^3 and 2 m^3** for **wet and dry waste**, respectively.

Required bin sizes for the study area is 1 m^3 and 2 m^3 for wet and dry waste, respectively. Calculated bin sizes are sufficient for collecting waste generated by 1230 people for more than a day due to 50% extra volume provided as FOS, assumed low waste density and higher per capita waste generation rate assumption. After calculating the required bin sizes, the number of bins required and the service area for a bin location were calculated for every ward of division-IV. The number of bins required for each ward are shown in Table 2 and sample calculation for finding the number of required bins and service area of each bin are shown below:

Ward-9

- Number of households = 7200 → Population = 7200 × 4.4 = 31,680
- Area of the Ward = 9,17,152 m^2
- Effective area = 6,14,000 m^2 (After reducing Empty/unhabituated area from the ward)
- Population density = 31,680/614000 = 0.051596 people/m^2 = 51,596 people/km^2
- A bin location serves 1230 people → In ward-9 in 0.024 km^2 area approx. 1230 people live i.e. 24,000 m^2
- Number of bins required for ward-9 = 31,680/1230 = **26 bins** (26 wet waste and 26 dry waste bin).

The number of bins required and the service area of each bin location for all 4 wards of division-IV are shown in Table 2. Service areas coverage of the bins in ward-9 is shown in Fig. 4. Solid waste generation in each of the service areas will be collected daily during the primary waste collection with the help of tippers and tricycles. After primary collection waste will be transferred to the respective bins for secondary storage and collection. Each service area shown in Fig. 4 is an area serving 2 bin locations. Providing a slightly large service area provide a choice for the waste collectors during the primary collection to transfer waste to the nearby located bin. This approach can really be helpful when the primary collection is done by a private partnership or by a local non-government organization (NGO's) like in the present case. Providing them with a slightly large area for waste collection may result in a more efficient primary waste collection because they have a better knowledge of road sizes, road conditions, and traffic problems in the local area. These parameters will directly affect the waste collection efficiency of the system. Also, monitoring primary collection in the urban areas of developing and under-developed nations via

Table 2 Number of bins required and service area of a bin location for each ward in division-IV

Ward No	Total population	Number of wet waste bin required (Size 1 m^3)	Number of dry waste bin required (Size 2 m^3)	Service area of a single bin location (in m^2)
9	31,680	26	26	24000
14	38,670	32	32	32037
15	33,770	28	28	37307
18	44,216	36	36	59752
Total	1,48,336	122	122	–

RS-GIS is very difficult due to the lack of road network and quality data availability in highly congested regions. So, it is better to provide a slightly larger service area rather than providing more smaller service areas.

After drawing service areas for the secondary bins, bins are located in the whole study area using Google Earth Pro. Bin location plan for division-IV is shown in Fig. 5, in which each green pin in the bin plan is a bin location representing 2 bins—wet waste bins (of size 1m^3) and dry waste bins (of size 2 m^3). To reduce the waste collection time and swift movement of the large collection vehicle, bins are located alongside the major roads of the city and clubbed together wherever possible. There are a total of 122 wet waste and 122 dry waste bins in the bin plan serving 148, 336 people which generates about 89 tonnes of MSW every day. All 244 bins are placed at the 76 locations in the study area in which four bins (2 wet waste and 2 dry waste) are provided at 39 locations and six bins are provided (3 wet waste and 3 dry waste bin) at 2 locations. Also, 2 bin locations (Bin 9–16 and Bin 18–16) are in the vicinity of the transfer stations, so at these locations, no extra bins are provided. So, now 120 bin locations need to be covered during the secondary waste collection.

3.2 *Vehicle Routing*

After designing a bin location plan for the study area solid waste collection routes were prepared using the manual vehicle routing approach. The number of maximum bins that can be covered in a single trip by a compactor truck was calculated before creating the collection routes. Detailed calculations for a single waste collection trip are shown below:

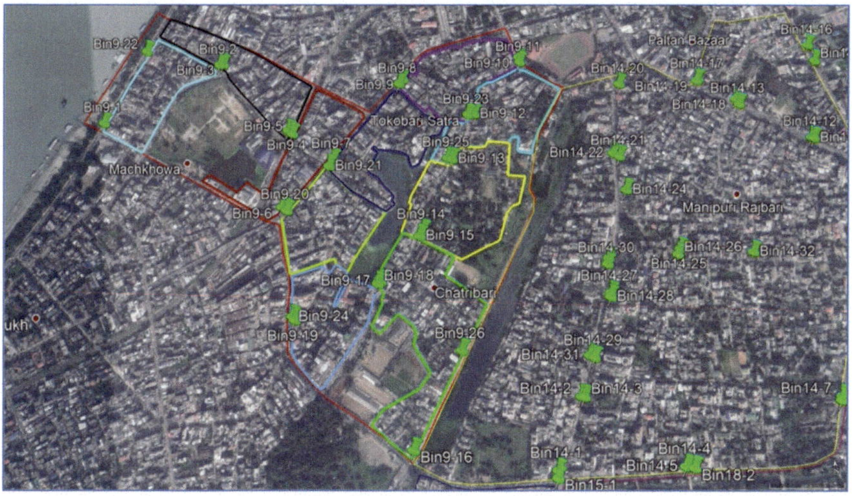

Fig. 4 Service areas for the bins in Ward-9

Wet waste bin covered in a single trip

Wet waste bin size = 1 m^3 bin; Density of the wet waste = 500 kg/m^3

After compaction = 1 m^3 × 500 kg/m^3 → 500 kg / 750 kg/m^3 → 0.67 m^3/ one bin location.

Bin points covered in a single trip → 14/0.67 → 20.89 → **21 locations** per trip*.

Dry waste bin covered in a single trip

Dry waste bin size = 2 m^3 bin; Density of the wet waste = 300 kg/m^3

After compaction = 2 m^3 × 300 kg/m^3 → 600 kg/750 kg/m^3 → 0.8 m^3/ one bin location.

Bin points covered in a single trip → 14/0.8 → 17.5 → **17 location** per trip.

In a single trip, a 14 m^3 truck can cover 21 wet waste bin locations or 17 dry waste bin locations. After knowing the maximum number of bins that can be covered in a single trip, collection routes were drawn for division-IV. All the collection routes were sketched in Google Earth Pro software for wet waste collection which needs to

Fig. 5 Bin location plans and all six-waste collection route for the division-IV (each green pin is bin point where 2 bins are placed for source segregation; Bin code is "Bin (ward no.) – (Bin number)". TS-1 and TF-Div-4 are the two-transfer station))

Table 3 Location and distance covered, and time taken by each route

Route number	Distance covered (in km)	Time taken (in min)	Location covered
1	8.00	151	21
2	5.92	142	21
3	5.80	142	21
4	4.76	137	20
5	6.13	143	21
6	4.43	135	16
Total	38.42*	930**	120

* Including returning distance to garage; ** Including break timings

be collected on a daily basis. The detailed calculation for creating routes is shown below (final results shown in Table 3):

Route-1: Total distance covered = 8 km Locations covered = 21 locations

Travelling time = 8 km × 60 min/h/20 km/h = 24 min × 1.5 (50% extra time) = 36 min.

Unloading time at transfer station = 10 min.

Bin unloading time = 5 min/locations × 21 locations = 105 min.

Total time taken = 36 + 10 + 105 min = 2 h 31 min.

By following same methodology collection time is calculated for other 5 routes.

Route-2: Total distance covered = 5.92 km	Location covered = 21 locations
Total time taken = 2 h 22 min	
Route-3: Total distance covered = 5.8 km	Location covered = 21 locations
Total time taken = 2 h 22 min	
Route-4: Total distance covered = 4.76 km	Location covered = 20 locations
Total time taken = 2 h 17 min	
Route-5: Total distance covered = 6.13 km	Location covered = 21 locations
Total time taken = 2 h 23 min	
Route-6: Total distance covered = 4.43 km	Location covered = 16 locations
Total time taken = 2 h 15 min	

Returning from the Transfer station (TS-Div-4): Distance = 3.38 km.

Total Time taken in returning to GMC garage = 10 min.

Time taken for all six routes = **14 h 20 min**.

Total time taken after including breaks = 14 h 20 min + 30 min + (2 × 15 min) = **15 h 20 min**

Total time taken in more than 8 h and less than 16 h. So, **2 trucks** are required.

Truck-1 first will follow Route-1, then Route-3, then Route-5 and finally go back to garage (Fig. 7).

Total distance covered by truck-1 = 8 + 5.8 + 6.13 + 3.38 = 23.31 km.

Total time taken by truck-1 = 7 h 26 min + 15 min + 15 min ~ **8 h.**

Truck-2 first will follow Route-2, then Route-4, then Route-6 and finally go back to garage.

Total distance covered by truck-2 = 5.92 + 4.76 + 4.43 + 3.38 = 18.49 km.

Total time taken by truck-2 = 7 h 4 min + 15 min + 15 min ~ **7 h 34 min.**

There are a total of six routes for covering 120 bins of the study area which required 15 h and 30 min. A single truck can maximum work for up to 8 h, so two trucks are required for the wet waste collection. Both the trucks will do the waste collection parallelly and after collection, both of them will end their trip at the transfer station-1. After completing a route truck will unload the waste at the transfer station. At the end of the day or on alternative days waste can be transferred from the transfer stations to the required destination (Landfill or Waste Treatment facility or MRF). Routes were drawn irrespective of ward-wise collection because it might result in more travel time and distance.

In the present study, only the secondary collection of the wet was taken into consideration. Using the same methodology collection routes can also be generated for dry waste collection. Due to high temperature and humidity in the region, wet waste has to be collected on an everyday basis, but recyclables do not require everyday collection. Recyclables can be collected on every day or every alternative day basis. But changing collection frequency will increase the size of the storage bin and reduce the number of locations that can be covered in a single trip by the same collection vehicle. So, a cost–benefit analysis is required for selecting everyday collection or alternate day collection of recyclables/dry waste. A detailed cost benefit analysis for selecting the collection frequency has already been published by our group for a different study area in which everyday collection was found more economical than other approaches (Singhal and Goel 2021a).

3.3 Time Allotment for Routing

The last task that remains for vehicle routing is the time allotment to the waste collection vehicles for covering various routes. However, time allotment is a complex task that requires good-quality traffic data. In the present study traffic conditions on the road were evaluated using Google Maps. Historical traffic conditions shown in Google Maps were used for identifying hourly traffic conditions on different weekdays. A snapshot of the traffic conditions in the study area was shown in Fig. 6.

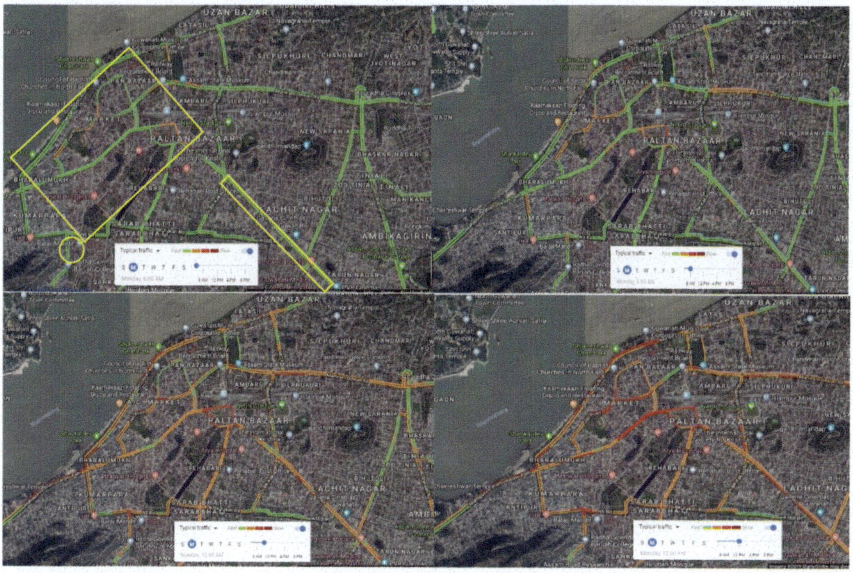

Fig. 6 Traffic conditions in division-IV at 6 am, 12 pm, 8 am and 10 am. In the first image busy traffic areas and junctions of the study area are marked by yellow colour. *Source* Google Maps; green → Fast/low traffic, orange → Moderate traffic, Red → Slow/Heavy traffic

From the images, it can be clearly noticed that traffic before 10 am is very light even in the busy market areas (marked area in Fig. 6). Between 10 am and 12 pm traffic is high in the market areas and major road junctions but varies from moderate to light traffic in the rest of the areas. So, waste collection in heavy traffic areas such as market areas will be done early in the morning and from light to moderate traffic areas can be collected around 12–1 pm. By doing this traffic congested and busy areas were collected early in the morning and low traffic-congested areas will be covered in the early afternoon. The total time required for the routing is a maximum of 8 h for each vehicle. So, a collection provided in this study was 5 am to 1 pm. Routes 1, 2, and 4 mostly lie in and around the heavy traffic area, so these routes are covered earliest in the day. Each of these routes will take a maximum of 2.5 h, so these routes can easily be covered before 10 am when traffic conditions are light to moderate. So, routing will be done: first truck will cover route-1 first, then route-3 and route-5 in the last (total time take is approximately 8 h); the second truck will cover the route-2 first, then route-4 and route-6 in the last (total time taken is 7 h 34 min). All the routes followed by each truck are illustrated in Fig. 7 with the total distance covered and time taken.

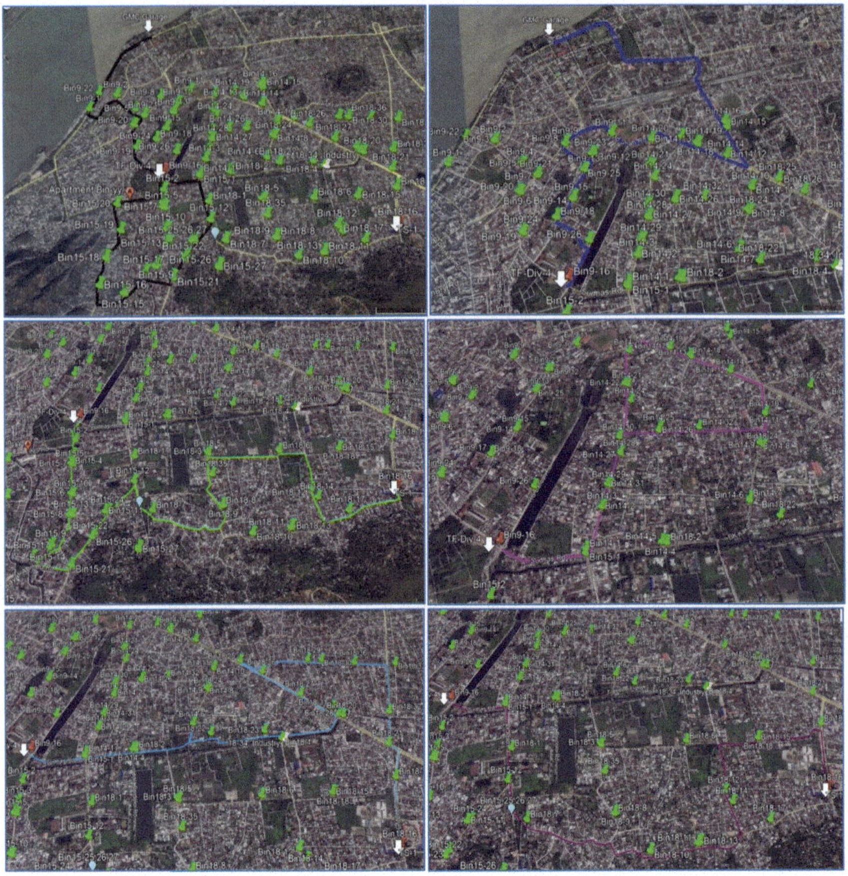

Fig. 7 Images of the left column are the routes covered by one truck and routes shown in right column will be collected by another truck (Total distance and time taken by 1st truck = 23.31 km and 8 h; by 2nd truck = 18.49 km and 7 h 34 min)

3.4 Benefits and Limitations of the Present Approach

The approach used in the present study is very simple and can easily be extended to other urban areas for designing WC&T systems with a minimal amount of data. This methodology could be really helpful for the local authorities which are struggling due to a lack of waste data or financial resources as the material cost required in the present approach is very low. The present approach might be a good choice when designing an MSWM system for a small city or towns or wards/sectors of a large city with different municipalities as it is less time-consuming and requires lower financial resources. Using historical and live traffic data from Google Maps, overall waste collection time and fuel consumption can be reduced significantly by avoiding heavy traffic areas during peak traffic hours. Due to less amount of data required and simple calculations, bin plans and collection routes can be developed and tested within a short amount of time. Collection time and routing efficiency can be further modified after pilot testing of the collection routes on the field. To further improve the collection efficiency, mathematical and linear optimization of collection routes can be done. Optimization of the collection routes should be implemented as it may result in enormous cost savings in fuel consumption and lead to fewer emissions from the waste collection fleet (Jaunich et al. 2016; Lella et al. 2017). However, it will increase the complexity of the present approach as mathematical optimization and programming require experienced and skilled supervision. Singhal and Goel (2021a, b, c) have tried to improve the efficiency of waste collection routes using Google Earth and linear programming which result in a 13% and 16% reduction in route length and collection time respectively.

Even though the present approach is a simple and effective way of designing the WC&T system, however, present approach has certain limitations which restricts its usability depending on certain conditions. The present approach is difficult to implement when routes have to be designed for the case in which there are more than 3–4 transfer stations or MRFs in the study area. In such cases, designing collection routes manually will be difficult and manually designed routes may lead to inadequate collection efficiency. For such case route optimization, mathematical or/and multicriteria decision modelling or Network analyst extension in Arc-GIS should be used (Das and Bhattacharyya 2015; Khan and Samadder 2016; Son and Louati 2016; Vu et al. 2019). Even though Google Earth is easy to use and contains a large data set of historical imagery, however planimetric accuracy of the images in Google Earth can vary significantly depending on the region/country (Goudarzi and Landry 2017; Pulighe et al. 2016). This may affect the accuracy of calculated population density data, designed bin service areas and collection routes. To minimize the error, georeferenced satellite images or satellite images from Landsat and MODIS should be used. However, requiring such images and performing analysis on them will require high internet-speed, high-quality computer systems and software such as Arc-GIS, Q-GIS, etc. Performing such complex RS and GIS analysis required skilled supervision and an expensive system which may be difficult to acquire for many local authorities even in developing nations (Singhal et al. 2021b). In such cases Google Earth can be effectively used for developing the WC&T plan as an error in position accuracy is usually within a few meters (Pulighe et al. 2016). Also, errors associated with calculated routing distance in Google Earth can easily be minimized during the pilot or practical testing of routes. So, even though the present approach has certain limitations in terms of feasibility and accuracy, still for many urban and rural areas, the present approach may still be valued as it requires the least number of resources.

4 Conclusion

The study mainly emphasizes the development of a waste storage and collection system using a simple approach, easily accessible RS and GIS tools and a minimal amount of external data. An efficient waste storage and collection system were developed in the present study by using Google Earth and Google Maps. The major outcomes of the study area:

1. Sizes of the wet waste and dry waste bins can be calculated by using basic population, waste composition, and per capita waste generation data, for efficient secondary waste storage.
2. The service area for every bin and bin location plan for the whole study area can be developed by using Google Earth and calculated population density data for each ward.
3. Six collection routes were designed for collecting 120 bins of the study area by using a manual routing approach and Google Earth. The total time required for the waste collection was 15 h 30 min, which can be achieved by using 2 collection vehicles simultaneously.
4. Using historical traffic data from Google Maps, a feasible time period was allotted for efficient waste collection to avoid heavy traffic areas during peak hours.

The approach used in the present study can be extended to other urban and rural areas in developing or underdeveloped nations for designing waste collection systems. This methodology could be really helpful for the local authorities which are struggling due to lack of data or financial resources as the material cost required in the present approach is very low as Google Earth and Google Maps both are freely available online.

Acknowledgements The authors would like to acknowledge Mr. Manojit Bujarbaruh (Executive Engineer, GMC, Guwahati) for his valuable contribution to the survey of the study area. The authors are grateful to Kartik Madhnure and Shubham Patil for their help in manuscript editing.

References

Agamuthu, P., Khidzir, K.M. and F.S. Hamid (2009). Drivers of sustainable waste management in Asia. *Waste Manag. Res. J.*, **27**: 625–633. https://doi.org/10.1177/0734242X09103191.

Annepu, R.K. (2012). Sustainable Solid Waste Management in India. MS Diss. https://doi.org/10.1007/978-981-4451-73-4.

Das, S. and B.K. Bhattacharyya (2015). Optimization of municipal solid waste collection and transportation routes. *Waste Manag.*, **43**: 9–18. https://doi.org/10.1016/j.wasman.2015.06.033.

Dutta, D. and S. Goel (2017). Applications of Remote Sensing and GIS in Solid Waste Management—A Review, *In:* Advances in Solid and Hazardous Waste Management. Springer International Publishing, Cham, pp. 133–151. https://doi.org/10.1007/978-3-319-57076-1_7.

Ghose, M.K., Dikshit, A.K. and S.K. Sharma (2006). A GIS based transportation model for solid waste disposal—A case study on Asansol municipality. *Waste Manag.*, **26**: 1287–1293. https://doi.org/10.1016/j.wasman.2005.09.022.

Goel, S. (2008). Municipal solid waste management (MSWM) in India: A critical review. *J. Environ. Sci. Eng.*, **50**: 319–328.

Goudarzi, M.A. and R.J. Landry (2017). Assessing horizontal positional accuracy of Google Earth imagery in the city of Montreal, Canada. *Geod. Cartogr.*, **43**: 56–65. https://doi.org/10.3846/20296991.2017.1330767.

Hannan, M.A., Akhtar, M., Begum, R.A., Basri, H., Hussain, A. and E. Scavino (2018). Capacitated vehicle-routing problem model for scheduled solid waste collection and route optimization using PSO algorithm. *Waste Manag.*, **71**: 31–41. https://doi.org/10.1016/j.wasman.2017.10.019

Hoornberg, D. and P. Bhada-Tata (2012). What a waste: A global review of solid waste management. Urban Development Series; knowledge papers no. 15. © World Bank, Washington, DC.

Hu, Q., Wu, W., Xia, T., Yu, Q., Yang, P., Li, Z. and Q. Song (2013). Exploring the use of google earth imagery and object-based methods in land use/cover mapping. *Remote Sens.*, **5**: 6026–6042. https://doi.org/10.3390/rs5116026.

Jaunich, M.K., Levis, J.W., DeCarolis, J.F., Gaston, E.V., Barlaz, M.A., Bartelt-Hunt, S.L., Jones, E.G., Hauser, L. and R. Jaikumar (2016). Characterization of municipal solid waste collection operations. *Resour. Conserv. Recycl.*, **114**: 92–102. https://doi.org/10.1016/j.resconrec.2016.07.012.

Khan, D. and S. Samadder (2016). Allocation of solid waste collection bins and route optimisation using geographical information system: A case study of Dhanbad City, India. *Waste Manag. Res. J.*, **34**: 666–676. https://doi.org/10.1177/0734242X16649679.

Kumar, L. and O. Mutanga (2018). Google Earth engine applications since inception: Usage, trends, and potential. *Remote Sens.*, **10**: 1509. https://doi.org/10.3390/rs10101509.

Laureri, F., Minciardi, R. and M. Robba (2016). An algorithm for the optimal collection of wet waste. *Waste Manag.*, **48**: 56–63. https://doi.org/10.1016/j.wasman.2015.09.020.

Lella, J., Mandla, V.R. and X. Zhu (2017). Solid waste collection/transport optimization and vegetation land cover estimation using Geographic Information System (GIS): A case study of a proposed smart-city. *Sustain. Cities Soc.*, **35**: 336–349. https://doi.org/10.1016/j.scs.2017.08.023.

Lin, H.-Y. and J.-J. Kao (2008). Subregion districting analysis for municipal solid waste collection privatization. *J. Air Waste Manage. Assoc.*, **58**: 104–111. https://doi.org/10.3155/1047-3289.58.1.104.

Lin, H.-Y., Tsai, Z.-P., Chen, G.-H. and J.-J. Kao (2011). A model for the implementation of a two-shift municipal solid waste and recyclable material collection plan that offers greater convenience to residents. *J. Air Waste Manage. Assoc.*, **61**: 55–62. https://doi.org/10.3155/1047-3289.61.1.55.

Liu, D. and S. Kenjeres (2017). Google-Earth based visualizations for environmental flows and pollutant dispersion in urban areas. *Int. J. Environ. Res. Public Health*, **14**(**3**): 247. https://doi.org/10.3390/ijerph14030247.

MacRae, G. (2012). Solid waste management in tropical Asia: What can we learn from Bali? *Waste Manag. Res.*, **30**: 72–79. https://doi.org/10.1177/0734242X10386138.

Malakahmad, A., Md Bakri, P., Md Mokhtar, M.R. and N. Khalil (2014). Solid waste collection routes optimization via GIS techniques in Ipoh city, Malaysia. *Procedia Eng.*, **77**: 20–27. https://doi.org/10.1016/j.proeng.2014.07.023.

O'Connor, D.L. (2013). Solid Waste Collection Vehicle Route Optimization for the City of Redlands , California. Master's Thesis, Univ. Redlands, pp. 1–53.

Pulighe, G., Baiocchi, V. and F. Lupia (2016). Horizontal accuracy assessment of very high resolution Google Earth images in the city of Rome, Italy. *Int. J. Digit. Earth*, **9**: 342–362. https://doi.org/10.1080/17538947.2015.1031716.

Sharholy, M., Ahmad, K., Mahmood, G. and R.C. Trivedi (2008). Municipal solid waste management in Indian cities—A review. *Waste Manag.*, **28**: 459–467. https://doi.org/10.1016/j.wasman.2007.02.008.

Singhal, A. and S. Goel (2021a). An Integrated Solid Waste Management (ISWM) Plan Using Google Earth and Linear Programming: A Case Study of Kharagpur City, West Bengal, *In:*

Treatment and Disposal of Solid and Hazardous Wastes. pp. 169–192. https://doi.org/10.1007/978-3-030-29643-8.

Singhal, A. and S. Goel (2021b). Spatio-temporal Analysis of Open Waste Dumping Sites Using Google Earth: A Case Study of Kharagpur City, India, *In:* Spatial Modeling and Assessment of Environmental Contaminants. Environmental Challenges and Solutions. pp. 137–151. https://doi.org/10.1007/978-3-030-63422-3_9.

Singhal, A. and S. Goel (2021c). Impact of sandstone quarrying on the health of quarry workers and local residents in Keru, Jodhpur, India. *In:* Treatment and Disposal of Solid and Hazardous Wastes. Springer International Publishing, pp. 103–125. https://doi.org/10.1007/978-3-030-29643-8.

Singhal, A. and S. Goel (2020). Environmental Impacts of Sandstone Quarrying and Its Waste: A Case Study of Jodhpur, India. *In:* 35th International Conference on Solid Waste Technology and Management. The Journal of Solid Waste Technology and Management, Pennsylvania, USA, pp. 326–338.

Singhal, A., Goel, S. and D. Sengupta (2020). Physicochemical and elemental analyses of sandstone quarrying wastes to assess their impact on soil properties. *J. Environ. Manage.*, **271**: 111011. https://doi.org/10.1016/j.jenvman.2020.111011.

Singhal, A., Gupta, A.K., Dubey, B. and M.M. Ghangrekar (2021a). Seasonal characterization of municipal solid waste for selecting feasible waste treatment technology for Guwahati city, India. *J. Air Waste Manage. Assoc.*, **72**(**2**): 1–14. https://doi.org/10.1080/10962247.2021.1980450.

Singhal, A., Madhnure, K., Dubey, B.K. and S. Goel (2021b). Application of Google Earth in solid waste management: For developing waste management plan and studying waste dumpsites. *In:* 18th International Symposium on Waste Management and Sustainable Landfilling. CISA Publisher, Cagliari, Italy.

Son, L.H. and A. Louati (2016). Modeling municipal solid waste collection: A generalized vehicle routing model with multiple transfer stations, gather sites and inhomogeneous vehicles in time windows. *Waste Manag.*, **52**: 34–49. https://doi.org/10.1016/j.wasman.2016.03.041.

Tavares, G., Zsigraiova, Z., Semiao, V. and M.G. Carvalho (2009). Optimisation of MSW collection routes for minimum fuel consumption using 3D GIS modelling. *Waste Manag.*, **29**: 1176–1185. https://doi.org/10.1016/j.wasman.2008.07.013.

Vecchi, T.P.B., Surco, D.F., Constantino, A.A., Steiner, M.T.A., Jorge, L.M.M., Ravagnani, M.A.S.S. and P.R. Paraíso (2016). A sequential approach for the optimization of truck routes for solid waste collection. *Process Saf. Environ. Prot.*, **102**: 238–250. https://doi.org/10.1016/j.psep.2016.03.014.

Vu, H.L., Bolingbroke, D., Ng, K.T.W. and B. Fallah (2019). Assessment of waste characteristics and their impact on GIS vehicle collection route optimization using ANN waste forecasts. *Waste Manag.*, **88**: 118–130. https://doi.org/10.1016/j.wasman.2019.03.037.

Worrell and Vesilind (2012). Solid Waste Engineering, 2nd ed. Cengage Learning.

Xue, W., Cao, K. and W. Li (2015). Municipal solid waste collection optimization in Singapore. *Appl. Geogr.*, **62**: 182–190. https://doi.org/10.1016/j.apgeog.2015.04.002.

Yu, L. and P. Gong (2012). Google Earth as a virtual globe tool for Earth science applications at the global scale: Progress and perspectives. *Int. J. Remote Sens.*, **33**: 3966–3986. https://doi.org/10.1080/01431161.2011.636081.

Chapter 10
Fundamentals of Microbiology

Naseeba Parveen, Tandra Mohanta, Deblina Dutta, and Sudha Goel

1 Introduction

While a large number of organisms are involved in the degradation of solid waste, microbes or microorganisms constitute the biggest group. Other than vermicomposting, all bioprocesses for solid waste treatment are designed based on microbes, mostly bacteria and fungi. Microbes are microscopic organisms that individually are too small to be seen by the naked human eye.

2 Classification of Living Organisms

All living organisms are made up of cells and can be broadly classified as prokaryotes or eukaryotes depending on differences in cell structure. All prokaryotes are unicellular, while eukaryotes can be unicellular or multicellular. Prokaryotes are also considered to be the most primitive organisms on an evolutionary basis.

N. Parveen
Civil Engineering Department, National Institute of Technologyy Mizoram, Aizawl, Mizoram 796012, India

T. Mohanta
Advanced Technology Development Center, Indian Institute of Technology Kharagpur, Kharagpur 721 302, India

D. Dutta
Department of Environmental Science and Engineering, School of Engineering and Sciences, SRM University-AP, Amaravati, Andhra Pradesh 522 240, India

S. Goel (✉)
Civil Engineering Department and School of Environmental Science and Engineering, Indian Institute of Technology Kharagpur, Kharagpur 721 302, India
e-mail: sudhagoel@civil.iitkgp.ac.in

S. Goel (ed.), *Advances in Solid and Hazardous Waste Management*,
https://doi.org/10.1007/978-3-031-49144-3_10

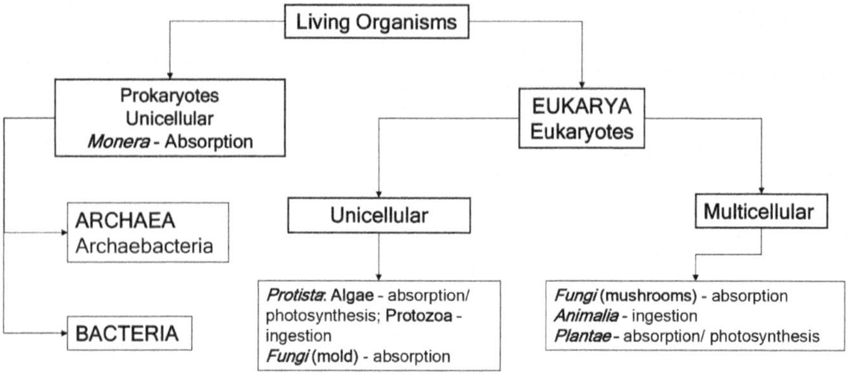

Fig. 1 Woese's and Whittaker's classification of microorganisms. Kingdoms in CAPITALS—rRNA based Woese's classification; Small case italics—Whittaker's classification

Fossil evidence suggests that prokaryotes were present 3.5 billion years ago while eukaryotes have been found around 1–1.9 billion years ago (Madigan et al. 2015).

Earlier when the number of species of living organisms that were known was limited, morphological differences were used as criteria to group organisms into two kingdoms—Plantae and Animalia. Later, R.H. Whittaker, in 1969, proposed the five kingdoms of classification based on their modes of nutrition and other criteria. These five kingdoms include Monera (all bacteria or prokaryotes, Protista, Fungi, Plantae and Animalia (Whittaker and Margulis 1978). Whittaker's classification of organisms and their modes of nutrition are shown in Fig. 1 and a comparison is provided in Table 1. At present, organisms have been classified based on criteria such as cellular structure, mode of nutrition, and genetic characteristics. Current textbooks use a 3-kingdom classification called Woese's classification (1990) based on similarities in ribosomal RNA: *Archaea*, Bacteria and Eukarya. *Archaea* bacteria are never pathogenic and all known bacterial pathogens belong to the domain Bacteria (Tortora et al. 2010).

However, other organisms have been studied by microbiologists and cannot be classified as living by conventional definitions. These include:

Viruses: Viruses are non-cellular, obligate, intracellular parasites. Outside the body of the host, they are non-living but inside the host, they show characteristics of living organisms such as:

a. Viruses have genetic material (DNA or RNA strands, but not both) that can be replicated and hence, can maintain heredity.
b. The genetic material is enclosed in a protein coat called capsid. The capsid is made of subunits called capsomeres. The protein coat allows the virus to survive as a free particle in the environment, i.e., in non-living form, which becomes viable or vegetative only after infecting a host cell.
c. The genetic material which is made of either DNA or RNA can mutate.

Viroids: Viroids are short strands of RNA without a protein coat.

Table 1 Some defining characteristics of organisms in Whittaker's 5 Kingdom classification

Characteristics	Kingdoms				
	Monera	Protista	Fungi	Plantae	Animalia
Type of cell	Prokaryotic	Eukaryotic	Eukaryotic	Eukaryotic	Eukaryotic
Cell wall	Non-cellulosic (polysaccharide + amino acid)	Present in some	Present (chitin)	Present (cellulose)	Absent
Nuclear envelope or nucleus	Absent	Present	Present	Present	Present
Cellular organisation	Unicellular	Unicellular	Unicellular or Multicellular with loose tissue	Tissue/ organ	Tissue/organ/organ systems
Mode of nutrition	Autotrophic: chemoautotrophic and photoautotrophic; Heterotrophic (saprophytic/parasitic)	Autotrophic: photoautotrophic and Heterotrophic	Heterotrophic (saprophytic/ parasitic)	Autotrophic: photoautotrophic	Heterotrophic: holozoic/saprophytic
Examples	All bacteria; *Escherichia coli, Pseudomonas aeruginos*a	Algae and protozoa	Unicellular—molds and yeasts Multicellular—mushrooms	All plants	All animals

Prions: These are infectious particles comprised entirely of proteins and do not possess any nucleic acid. They were discovered by SB Prusiner in 1997 and are the cause of several neurological diseases in humans and animals such as mad cow disease. Prions are the most resistant forms known at present and are capable of surviving conventional sterilization methods.

2.1 *Differences Between Prokaryotes and Eukaryotes*

Prokaryotes and eukaryotes can be differentiated based on several characteristics and these differences are summarized in Table 2.

Table 2 Principal differences between prokaryotic and eukaryotic cells

Characteristics	Prokaryotes	Eukaryotes
Size of cell	Typically 0.2–2.0 μm in diameter	Typically 10–100 μm in diameter
Nucleus	No nuclear membrane or nucleoli	True nucleus, consisting of nuclear membrane and nucleoli
Chromosome (DNA)	Single circular chromosome, lacks histones	Multiple linear chromosomes with histones
Membrane enclosed organelles	Absent	Present
Flagella	Consist of two protein building blocks	Complex, consist of multiple microtubules
Glycocalyx	Present as a capsule or slime layer	Present in some cells that lack a cell wall
Cell wall	Usually present, chemically complex (typical bacterial cell wall includes peptidoglycan)	When present, chemically simple
Plasma membrane	No carbohydrates and generally lacks sterols	Sterols and carbohydrates that serve as receptors present
Cytoplasm	No cytoskeleton or cytoplasmic streaming	Cytoskeleton, cytoplasmic streaming may be present
Ribosomes	Smaller size (70s)	Larger size (80s), smaller size (70s) in organelles such as mitochondria
Cell division	Binary fission	Mitosis
Sexual recombination	No meiosis, transfer of DNA fragments only (plasmid transfer)	Involves meiosis

2.2 *Functions of Prokaryotic Structures*

- Plasma membrane: It is a selectively permeable barrier that serves as the mechanical boundary of the cell through which nutrients and wastes are transported. It is the location of many metabolic processes (respiration, photosynthesis), and allows the detection of environmental cues for chemotaxis.
- Gas vesicles or vacuoles: these provide the cell with buoyancy for floating in aquatic environments; they are very important for photosynthetic organisms like cyanobacteria, algae and green and purple photosynthetic bacteria that need to maximize their utilization of sunlight.
- Ribosomes: ribosomes are sites for protein synthesis.
- Inclusion bodies: they are organelles for the storage of carbon, phosphate, and other substances
- Nucleoid: since prokaryotes have no true nucleus, the cell volume occupied by genetic material, i.e., the single strand of DNA is called the nucleoid.
- Periplasmic space and periplasm: this is the space between the outer membrane and the cytoplasmic membrane which contains a gel-like fluid containing hydrolytic enzymes and binding proteins for nutrient processing and uptake (Tortora et al. 2010).
- Cell wall: the cell wall gives bacteria shape and protection.
- Capsule and slime layers: These help the cells to adhere to surfaces and provide resistance to phagocytosis.
- Fimbriae and pili: these help the cell to attach to surfaces. Pili help in bacterial mating (conjugation).
- Flagella are responsible for the ability of the cell to move in its environment.
- Endospore: Endospores are the dormant form of bacterial cells that allows them to survive under harsh environmental conditions; endospores have been known to survive from 34 years to 250 million years (Madigan et al. 2003) and are second only to prions in terms of resistance to chemical biocides (Tortora et al. 2010).

2.3 *Functions of Eukaryotic Organelles*

- Plasma membrane: It is a selectively permeable barrier that serves as a mechanical boundary of the cell through which nutrients and wastes are transported. It also mediates cell–cell interactions and adhesion to surfaces, and the secretion of wastes and extracellular materials including enzymes.
- Cytoplasmic matrix: The cytoplasm is the fluid contained within the plasma membrane that provides a medium for metabolic processes to take place and contains the organelles and other materials needed for the cell's growth and survival.
- Microfilaments, intermediate filaments and microtubules: These are responsible for cell structure and movements and form the cytoskeleton.

- Endoplasmic reticulum (ER): ER is an extensive network of membranous sacs or tubules called cisterns and is continuous with the nuclear membrane (Tortora et al. 2010). ER can be of two types: rough or smooth ER. Rough ER has ribosomes attached to it and is the site of protein and membrane molecule synthesis. Smooth ER has no ribosomes attached to it but has unique enzymes which allow the synthesis of fats, lipids and hormones like estrogen and testosterone.
- Golgi apparatus or Golgi complex: The first step in the transport of proteins from the ribosomes to other parts of the cell is through Golgi complexes which are stacks of 'cisterns' (Tortora et al. 2010). Proteins at the Golgi complex are moved through secretory or storage vesicles.
- Lysosomes: These are major storage vesicles that contain several powerful enzymes that are essential for intracellular digestion. In humans, white blood cells which are an important part of the body's defense mechanism against bacteria and other 'foreign' particles or organisms, have large numbers of lysosomes (Tortora et al. 2010).
- Mitochondria: Also known as powerhouse of the cell, where energy production occurs through the use of the tricarboxylic acid cycle, electron transport, oxidative phosphorylation, and various other pathways. It is capable of multiplying on its own by binary fission (Tortora et al. 2010).
- Chloroplasts: These are present in algae and green plants and are the site of photosynthesis. These are also capable of multiplying on their own by binary fission (Tortora et al. 2010).
- Nucleus: This is the repository for genetic information, i.e., DNA, and is the control centre for an individual cell.
- Nucleolus: Site of ribosomal RNA synthesis, and ribosome construction.
- Cell wall: The cell wall strengthens and gives shape to the cell.
- Cilia and flagella: These are responsible for cellular movement
- Vacuoles: Temporary storage and transport, digestion (food vacuoles), water balance (contractile vacuoles).

3 Enumeration and Visualization of Microbes

The small size of microbes makes their enumeration and visualization a major challenge in microbiology. Several methods and techniques have been developed in the last 400–500 years to observe, enumerate and visualize microbes. One of the oldest methods for the enumeration of viable or living cells is plating them on nutrient agar plates. This can be done using pour plate or spread plate methods and different types of media are now commercially available for the identification of individual species (Goel 2019; Madigan et al. 2015; Tortora et al. 2010). Solid waste samples are likely to have very high microbial content and the microbes in these samples can be brought into an aqueous medium by mixing with sterile buffer solutions and diluted to a level where the colonies on the plates can be counted in terms of colony forming units (cfu). These standard methods are described in most textbooks and reference books (APHA et al. 2005). The limitation of these methods is that only colonies can be counted and it is assumed that each colony represents a single cell. Plating methods do not allow us to visualize individual cells or their organelles. For visualizing and enumerating individual cells or their organelles, we can use light or electron microscopy. The basics of microscopy are described here briefly.

3.1 Light Microscopy

A light microscope uses a light source and magnifying lenses to observe microbes. The magnification of the light microscope varies depending on the type and number of lenses used. A simple light microscope uses two convex lenses while a compound light microscope uses a combination of three lenses (the condenser, objective and eyepiece or ocular lenses) resulting in higher magnification than the former. The working principle of a compound light microscope and a photograph of a compound light microscope with its components marked is shown in Fig. 2.

Resolution is the ability to distinguish two points at a specified distance apart. The resolution of the human eye is approximately 0.2 mm while that of light microscopes can go down to 0.2 μm which is 200 nm, depending on the type of technique used. The most used types of microscopes for microbial visualization are explained below.

The minimum distance (d) between two points that can be distinguished with a light microscope can be calculated using the Abbé equation.

$$d = 0.5\,\lambda/\mathrm{NA} = 0.5\,\lambda/n\,\mathrm{Sin}\,\theta$$

where NA = numerical aperture of the objective lens = n Sin θ, n = refractive index of the medium (between cover slip and objective lens) through which light passes, θ = one half of the objective lens' aperture angle, and λ is the wavelength of light.[1]

Light has to pass through the glass slide on which the specimen is mounted, then through the specimen and cover slip and finally through the medium before it enters the objective lens. The medium can be either air, water or oil. When air is the medium through which light passes, some of the light is lost and does not enter the objective lens due to the difference in the refractive index of air or vacuum which is 1.0 and that of glass is 1.5. Immersion oil is specially manufactured to have a refractive index close to that of glass, i.e., n = 1.5 so that light lost by refraction is minimized. The maximum magnification of a light microscope that uses visible light ranges from 1000 to 1500×. Standard compound light microscopes have 10× or 20×, 40×, and 100× objective lenses and the eyepiece has a magnification of 10×. The 100× objective lens is an oil immersion lens. In the absence of oil, water can be used as it has a refractive index of n = 1.33 and will give better results than air. The

[1] https://zeiss-campus.magnet.fsu.edu/articles/basics/resolution.html.

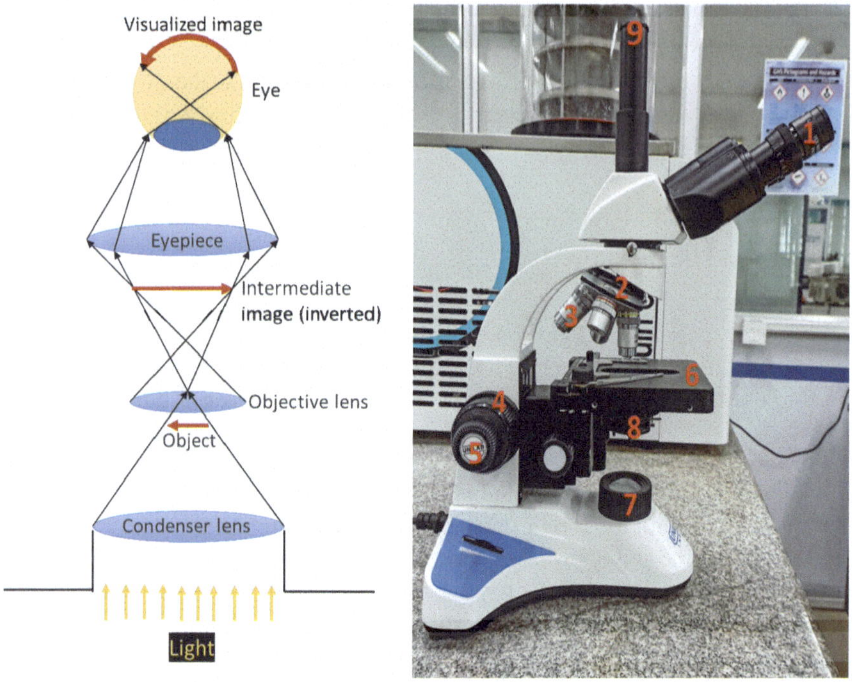

Fig. 2 Schematic of the working principle of a light microscope and a photograph of Unilab India® compound brightfield microscope; Parts marked (1) Eyepiece; (2) objective turret; (3) objective lenses of different magnification; (4) coarse adjuster; (5) fine adjuster; (6) specimen stage; (7) light source; (8) diaphragm and condenser; (9) camera point

highest resolution possible with a compound light microscope is around 0.2 μm and it is difficult to distinguish objects smaller than this.

Different light microscopes are available commercially and their working principles are described here (Tortora et al. 2010):

Bright-field microscope: In a bright-field microscope, light passes through the specimen and differences in opacity or colour allow the object to be seen. Large cells like *Paramecium* and algal cells which contain chlorophyll can be seen directly under the microscope. Smaller cells such as bacterial cells that are colourless need to be stained so that there is a contrast between the specimen and the surrounding environment.

Dark-field microscope: An opaque disk is placed between the light source and the condenser lens so that no light passes through the specimen directly. Instead, light reaches the specimen from the sides of the disk and is reflected off the specimen resulting in a light image against a dark background.

Phase-contrast microscope: In another modification to the bright field microscope, an annular diaphragm is placed in front of the light source. A fraction of the light from the source passes through the annular diaphragm and then through the condenser lens. This light is then brought to focus on the specimen. The organelles in the cell have different refractive indices. The path length of the light beams will be altered depending on the refractive indices of the different organelles. So, the refracted light as well as the transmitted light will pass through the objective lens and from there through a diffraction plate after which it will reach the eye lens and the eye. Thus, the differences in refractive indices of the different parts of the cell are used to create an image and details of the internal structures can be seen.

Fluorescence microscopy: Fluorescence microscopy is a relatively new development in light microscopy where fluorophores, i.e., compounds that fluoresce, are used to dye the specimen and image it with a fluorescence microscope. Fluorescence is a phenomenon where the light emitted by the object is at a different wavelength compared to the incident light. Some compounds like chlorophyll are naturally fluorescent which means they look green under normal sunlight and have a different colour, red in the case of chlorophyll, under UV light. The same phenomenon can be used to visualize other types of cells, which do not have naturally fluorescent compounds like chlorophyll. Fluorescent dyes like acridine orange are generally used to detect bacteria and other microbes.

Images obtained using each of these methods are shown in Fig. 3.

Cell Staining for Microscopy

Most unicellular microbes are not visible without staining. Cell staining provides a contrast against the background, making them visible under the microscope. By using different stains, cell organelles such as the nucleus or cell membrane or whole cell can be stained. Staining can be (i) simple staining which uses a single dye to stain the cell directly (simple direct staining) or to stain the background of the cell (simple negative staining) or (ii) differential staining where a combination of dyes is used to distinguish between organisms based on their cell characteristics. An example of a simple stain is acetocarmine used here for observing mitosis in the cells from the plant's root tips. Microphotographs of cells with different staining methods are shown in Fig. 4.

The most common differential staining method in microbiology is the Gram stain. Gram-positive bacteria are stained mid to dark purple as their cell walls have a single plasma membrane enclosed by thick, cross-linked peptidoglycan. Gram-negative bacteria are stained pink to red as their cell walls have a thin layer of peptidoglycan in the periplasmic space within their inner and outer lipid plasma membranes (Fig. 4b–d).

3.2 Electron Microscopy

Light microscopes can be used to visualize Bacteria and *Archaea*; however, viruses cannot be observed with them. For visualizing viruses and the internal details of prokaryotic cells, electron microscopes are used. Electron microscopes use a beam

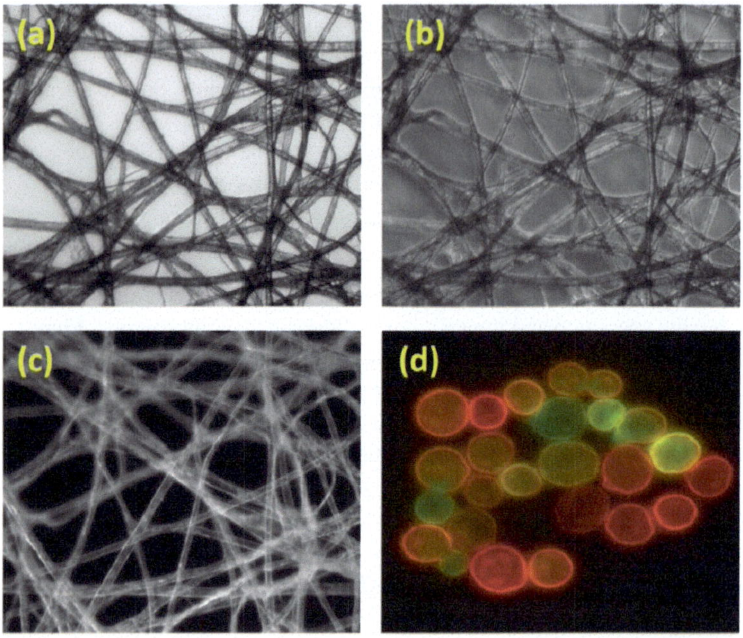

Fig. 3 Examples of light microscopic techniques used to generate contrast in a sample of tissue paper. **a** Bright-field illumination; sample contrast comes from the absorbance of light by fibers in the tissue paper; **b** Phase contrast illumination; sample contrast comes from the interference of different path lengths of light through the sample; **c** Dark field illumination; sample contrast comes from light scattered or reflected by the sample; **d** Yeast (S. cerevisiae) cell membrane visualized by fluorescent markers. *Source* By Zephyris—Own work, https://commons.wikimedia.org/w/index.php?curid=10762662; By Masur—Own work, https://commons.wikimedia.org/w/index.php?curid=905559

of electrons instead of light and electromagnets instead of lenses. Electrons are characterized by wavelengths that are 100,000× shorter than visible light wavelengths. This increases the 'd' value in the Abbe equation and increases the resolution of electron microscopes to 2 nm with a magnification of around 150,000×. Two types of electron microscopes are common and are described here.

Transmission electron microscope (TEM): In TEM, electron beams of TEM pass through the lighter areas of the specimen and are scattered by the denser areas of the specimen resulting in a detailed image of the cell structure. Sample preparation for TEM is extensive as thin sheets of 20–60 nm are needed that will allow the electron beam to pass through. Heavy metals may be used for staining the specimen as well.

Scanning electron microscope (SEM): SEM is used for studying the surface features of the specimen and can be used for enumerating the number of cells in a sample. Platinum or gold-coated specimens scatter the electron beam, and an image is obtained from the detector that collects the scattered electrons (secondary electrons

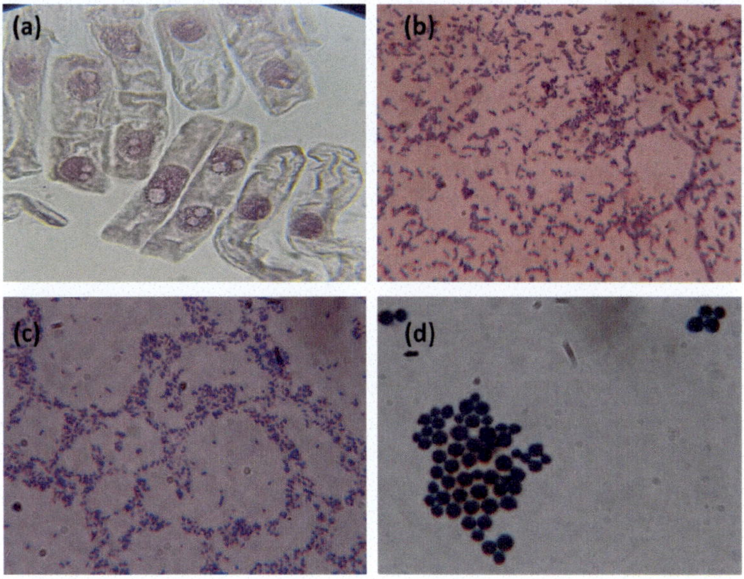

Fig. 4 **a** *Allium cepa* (onion) cells stained with acetocarmine; **b**, **c** Gram negative and Gram positive bacteria in a wastewater sample; **d** Gram positive bacteria (*Staphylococcus*) photographed with a brightfield microscope at 1000 × magnification

and backscattered electrons) to generate an image.[2] Secondary electrons are used for showing the morphology and topography of the specimen and backscattered electrons are used for producing contrasts.

3.3 Scanned Probe Microscopy

Scanned Probe Microscopy (SPM) allows magnification up to 100,000,000× and a resolution up to 0.1 nm, which helps to visualize down to the atomic level. SPM uses an extremely sharp probe to scan the specimen surface and photographs the atomic level details to form extensive 3D images. SPM can be used for studying liquids as well, which widens its application to biomolecules. Two types of scanned probe microscopes are available: Scanning Tunneling Microscope (STM) and Atomic Force Microscope (AFM).

Scanning tunneling microscope (STM): The tungsten probe of STM is one atom thick and maintains a constant voltage, allowing it to scan the surface of the specimen. STM works based on many principles including quantum mechanical tunneling and

---

[2] https://www.umassmed.edu/cemf/whatisem/.

the piezoelectric effect. It can be used to study the chemical and magnetic properties of materials and the temperature variations within a cell.

Atomic force microscope (AFM): AFM was proposed as an alternative to STM in 1986 (Binnig et al. 1986). AFM is a combination of both STM and the stylus profilometer. Unlike STM, AFM has a thin diamond probe that is in contact with the specimen surface allowing it to scan non-conductive and semi-conductive specimens.

4 Microbial Metabolism

All microbes take up dissolved compounds from their environment, and use them for generating energy for maintenance and creating new biomass. During these metabolic activities of energy and biomass generation, waste and toxic substances are excreted into the environment as shown in Fig. 5.

Substrates

Both organic and inorganic compounds can serve as substrates for biomass and energy generation. Inorganic compounds include gases like O_2, CO_2 and dissolved compounds such as Fe^{2+}, NO_2^-, H_2S, H_2, and S.

Nutrients

Bacterial cells are generally assumed to be 70–90% water and the total weight of a single bacterial cell is assumed to be 1 picogram (pg, i.e., 10^{-12} g). Out of the 96 elements found in nature, only 26 are used by living organisms (Tortora et al. 2010). Analysis of microbial cell composition (Table 3) shows that over 95% of the dry weight of the cell is made up of a few major elements: carbon (C), oxygen (O), hydrogen (H), nitrogen (N), sulphur (S), sodium (Na), phosphorus (P) and potassium (K) (Metcalf and Eddy 2003). Carbon constitutes more than 50% while N comprises about 13–14% of the dry weight of the cell. The first six nutrients or elements (C, O, H, N, S, and P) are components of carbohydrates, lipids, proteins, and nucleic acids. Carbohydrates and lipids are mainly composed of C, H and O while proteins additionally contain N and S. Nucleic acids, besides the first five elements, have a large component of P. The remaining two macronutrients exist in the cell as

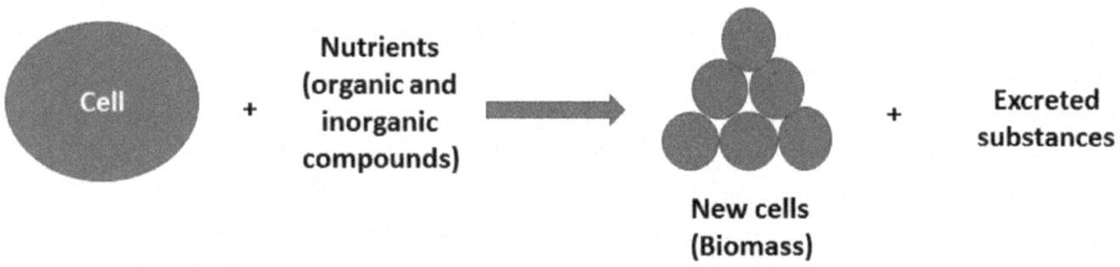

Fig. 5 Schematic representation of metabolism in microbes

Table 3 Typical composition of bacterial cells (Percentage of dry mass) (Metcalf and Eddy 2003)

Element	Range	Typical	Nutrient type
Carbon	45–55	50	Macronutrients
Oxygen	16–22	20	
Nitrogen	12–16	14	
Hydrogen	7–10	8	
Phosphorous	2–5	3	
Sulphur	0.8–1.5	1	
Potassium	0.8–1.5	1	
Sodium	0.5–2.0	1	
Calcium	0.4–0.7	0.5	Micronutrients
Magnesium	0.4–0.7	0.5	
Chlorine	0.4–0.7	0.5	
Iron	0.1–0.4	0.2	
All others	0.2–0.5	0.3	

cations and play a variety of roles. Sodium (Na) is an essential nutrient for some organisms, especially marine organisms. Potassium (K^+) is required for the activity of different enzymes, including those involved in protein synthesis. These elements which comprise >1% of the dry weight of the cell are termed macronutrients.

All organisms including microorganisms require several micronutrients or trace elements. Micronutrients are defined here as those that constitute < 1% of the dry weight of the cell. These include calcium, magnesium, iron, manganese, zinc, cobalt, molybdenum, nickel, and copper, and are needed by most cells. Calcium (Ca^{2+}), among other functions, contributes to the heat resistance of bacterial endospores. Magnesium (Mg^{2+}) serves as a cofactor for many enzymes, complexes with ATP, and stabilizes ribosomes and cell membranes. Iron (Fe^{2+} and Fe^{3+}) is a part of cytochromes and a cofactor for enzymes and electron-carrying proteins (Madigan et al. 2015).

4.1 Classification of Organisms Based on Food and Energy Sources

All organisms require nutrients and energy for growth and reproduction. There are two ways of classifying organisms. Organisms can be classified based on either carbon (C)-sources or energy sources.

Based on Carbon sources, organisms can be either autotrophs (meaning self-feeding) or heterotrophs (meaning other-feeding). Autotrophs convert inorganic C (mainly carbon dioxide) to organic carbon or biomass, while heterotrophs convert organic C (mainly other organisms or dead detrital matter) to new forms of organic carbon or new biomass. Plants and phytoplankton are examples of autotrophs while all organisms that consume other living organisms are examples of heterotrophs.

Based on energy sources, organisms can be either phototrophs or lithotrophs. Phototrophs derive their energy from light and include all photosynthetic organisms like plants, and algae. Many pigmented bacteria like cyanobacteria are photoautotrophs. Chemotrophs derive their energy from the reduction of inorganic compounds like Fe(III), Mn(IV), sulphate, and nitrate or the oxidation of hydrogen sulphide, sulphur, ammonia, nitrite ions, hydrogen, and ferrous ion.

Therefore, based on the above two categories, an organism can belong to any one of the following four groups (Tortora et al. 2010):

1. **Photoautotrophs**: They include plants, algae and some bacteria. They use carbon dioxide as their source of carbon and light as their source of energy. These organisms may be pigmented or non-pigmented, and they may be oxygenic or anoxygenic. Green and purple bacteria are well-known examples of photoautotrophs.
2. **Chemoautotrophs**: They include only a few species of bacteria. They use carbon dioxide as the carbon source and energy is obtained from the oxidation of inorganic compounds (hydrogen sulphide, sulphur, ammonia, nitrite ions, hydrogen, and ferrous ions) is the energy source and carbon dioxide is the carbon source. Nitrifying bacteria are found in surface water bodies with significant concentrations of ammonia as well as in wastewater treatment and are the best-known examples of chemo- or lithoautotrophs.
3. **Photoheterotrophs**: They include a small group of bacteria. They use simple organic compounds such as alcohols, fatty acids, organic acids and carbohydrates as their carbon and light as their energy source. Green and purple non-sulphur bacteria are examples of photoheterotrophs.
4. **Chemoheterotrophs**: They include protozoa, fungi, animals, and most bacteria. They utilize organic compounds as their carbon source and obtain energy from oxidation-reduction of organic compounds.

Heterotrophs are of three types based on their nutritional methods:

1. **Saprophytes**: They live off of dead organic matter and obtain nutrients through absorption. The best examples are fungi like *Mucor* and *Rhizobium.*
2. **Holozoic**: They ingest solid food or entire particles or other organisms, e.g., *Amoeba.*
3. **Parasites**: They obtain their nutrients from a living host, e.g., all disease-causing organisms or pathogens can be categorized as parasites.

Oxygen may or may not be required for metabolism. An organism is classified as aerobic or anaerobic based on its oxygen requirements. Aerobes require oxygen for metabolic activity and those organisms that do not need oxygen for their metabolic activities are called anaerobes.

4.2 *Metabolic Pathways*

Chemotrophic organisms (such as those found in Solid waste treatment processes that can be autotrophic or heterotrophic) obtain their energy and Carbon by following any of the metabolic pathways listed below. Each of these pathways is a series of oxidation-reduction reactions through which electrons are transported from an electron donor (ED) to a terminal electron acceptor (TEA) whilst generating ATP in the process.

- Aerobic respiration
- Anaerobic respiration and
- Fermentation.

Each of these three pathways is summarized here in terms of the terminal electron acceptor (TEA), electron donor (ED), ATP yield for glucose and mode of ATP generation. Complete or partial oxidation (no reduction) of organic compounds occurs only in the respiration pathways while the same organic compound is partially oxidized and partially reduced by fermentation. Many bacterial species are capable of switching from aerobic to anaerobic respiration depending on the availability of terminal electron acceptors (TEA).

Aerobic Respiration

- TEA: molecular oxygen; ED: can be any organic compound (e.g., glucose) or inorganic compound (e.g., ammonia or nitrite)
- ATP yield/glucose molecule = 38 (highest possible yield)
- Mode of ATP generation: substrate-level and oxidative phosphorylation.

Anaerobic Respiration

- TEA: inorganic compounds except O_2; examples are nitrate, sulfate, carbonates, Fe(III), and Mn(IV); ED: can be any organic or inorganic compound
- ATP yield/glucose molecule ≥ 2 to <38 (yield depends on the combination of TEA and ED)
- Mode of ATP generation: substrate-level and oxidative phosphorylation.

Fermentation

- TEA and ED: same organic compound can serve as both ED and TEA
- ATP yield/glucose molecule = 2
- Mode of ATP generation: substrate-level phosphorylation
- For example, glucose is converted to ethanol (reduced) + CO_2 (oxidized).

Examples of oxidation-reduction reactions for acetate with different TEAs are provided in 'Basic concepts from biochemistry' (Sawyer et al. 2003).

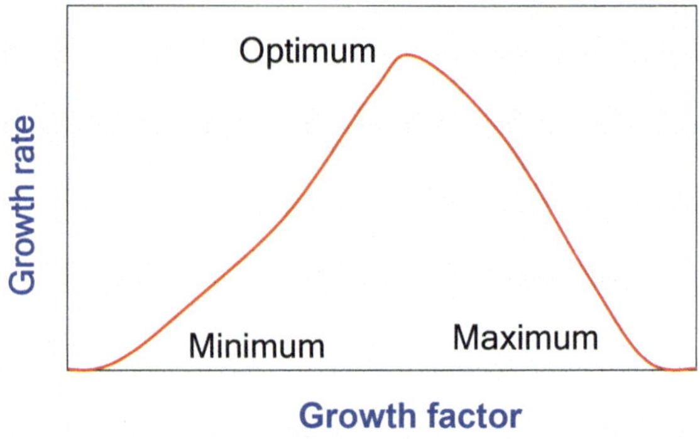

Fig. 6 Microbial growth rate in response to changes in specific growth or environmental factors

4.3 *Environmental Conditions*

Various environmental factors or conditions can affect the extent and rates of growth of microbes. Major factors are discussed here. In general, there are three cardinal points for every growth or environmental factor (Fig. 6).

Minimum level or concentration at which no growth occurs.

Maximum level or concentration at which no growth occurs, and Optimum level or concentration for any environmental factor at which microbial growth rates are highest.

1. **Nutrient availability**

Organisms require both organic as well as inorganic nutrients for creating new biomass. The law of the minimum defines a 'limiting nutrient' and it is important to note that the rate and extent of growth of any organism are limited by the availability of this 'limiting nutrient', also called a substrate.

Law of the minimum

This fundamental principle that applies to the growth of all living organisms was first formulated in agricultural science by Carl Sprengel in 1828 and later popularized by Justus von Liebig.[3]

Any element that is least available relative to the requirements of an organism is defined as the limiting nutrient since the growth of the organism is directly proportionate to the concentration of the limiting nutrient.

In general, any one of the macronutrients is often considered to be the limiting nutrient in an aquatic system. However, *it is important to note that any element (macro- or micro-nutrient) can be the limiting nutrient for the growth of an organism.*

Besides macro- and micronutrients organisms require various other growth factors for proper growth and development. These growth factors include amino acids, and vitamins and are compounds that cannot be synthesized by the organism and therefore, must be supplied externally.

[3] Wikipedia (2015) https://en.wikipedia.org/wiki/Liebig%27s_law_of_the_minimum.

2. **Temperature**

Temperature plays an important role in the growth and development of microbes. Higher temperatures speed up chemical reactions. For this reason, the growth rates of microbes tend to increase with an increase in temperature but only up to a certain point after which cells begin to die. This is mainly because, at very high temperatures, denaturation of proteins (enzymes) and nucleic acids takes place resulting in a loss of metabolic capabilities.

Cardinal temperatures: every organism can be characterized by three temperatures (Madigan et al. 2015):

- **Minimum temperature** below which no growth occurs
- **Optimum temperature** at which the fastest growth occurs
- **Maximum temperature** above which no growth occurs.

Depending on the adaptation of microorganisms to different temperatures, they are classified into four categories as noted below:

1. **Psychrophiles**—optimum temperature is typically 4 °C and maximum temperature is around 10 °C or lower; they are able to grow at temperatures below 0 °C.
2. **Mesophiles**—These organisms have a temperature range of 10–50 °C with an optimum temperature of 39 °C and a minimum temperature of around 10 °C.
3. **Thermophiles**—These organisms have a temperature range of 40–70 °C with an optimum temperature of 60 °C.
4. **Hyperthermophiles**—These organisms have a temperature range of 65–115 °C with optimum temperatures of 80 °C or higher. These organisms are mostly *Archaea* and are found in hot springs and deep-sea hydrothermal vents where the temperatures can exceed the boiling point of water.

The upper-temperature limits for the growth of different living organisms are shown in Table 4 (Madigan et al. 2015).

3. **pH**

The range of pH over which an organism grows is defined by **three cardinal points**:

- **Minimum pH** below which the organism cannot grow,
- **Maximum pH** above which the organism cannot grow, and
- **Optimum** Error! Bookmark not defined. **pH** at which the organism grows best.

Depending on their growth at different pH, microorganisms are classified as:

- **Acidophiles**—optimal acidic pH range of 1–5.5
- **Neutrophiles**—optimal pH range of 5.5–8
- **Alkaliphiles**—optimal pH range of 8.5–11.5
- **Extreme alkaliphiles**—optimum pH range of 10 or greater.

Table 4 Presently known upper temperature limits for growth of living organisms

Group	Upper temperature limits
Animals	
• Fish and other aquatic vertebrates • Insects • Ostracods (crustaceans)	38 45–50 49–50
Plants	
• Vascular plants • Mosses	45 50
Eukaryotic microorganisms	
• Protozoa • Algae • Fungi**Error! Bookmark not defined.**	56 55–60 60–62
Prokaryotes	
*Bacteria***Error! Bookmark not defined.**	
• Cyanobacteria • Anoxygenic phototrophs • Chemoorganotrophic/chemolithotrophic bacteria • **Error! Bookmark not defined.**	70–74 70–73 95
Archaea	
• Chemoorganotrophic/chemolithotrophic *archaea*	113

In the preparation and use of culture media, maintenance of optimum pH is required for the growth of a desired organism. Accordingly, buffer solutions should be incorporated to maintain the pH of the medium in response to bacterial waste products that accumulate during growth. Many pathogenic bacteria exhibit a relatively narrow range of pH over which they grow. Most diagnostic media for the growth and identification of human pathogens have a pH near 7.

It is important to note that most bacteria are neutrophiles (exceptions include some bacteria in acid mine drainage which have an optimum pH of 1–3) and most fungi prefer slightly acidic conditions (pH 4–6).

4. **Oxygen**

Oxygen is readily available in aqueous systems as dissolved oxygen or when H_2O is a reactant in biochemical reactions. However, microorganisms display a wide range of responses to molecular oxygen O_2. Depending on the requirements of oxygen, organisms are classified into two broad categories: aerobes and anaerobes.

Aerobes are those organisms that require oxygen for growth. Aerobes are further classified as obligate or facultative.

1. **Obligate aerobes** require O_2 for growth; they use O_2 as their terminal electron acceptor (TEA) in aerobic respiration.
2. **Facultative aerobes** (or **facultative anaerobes**) are organisms that can switch between aerobic and anaerobic metabolism depending on the availability of

oxygen. Under aerobic conditions (in the presence of O_2) they grow by aerobic respiration, but in the absence of O_2 they can switch to fermentation or anaerobic respiration.

3. **Microaerophiles**—these bacteria don't grow at normal atmospheric oxygen (21%), but require small amounts of oxygen for growth (2–10%).

Anaerobes do not require oxygen for growth. They are also classified as obligate or facultative.

1. **Obligate anaerobes** (occasionally called **aerophobes**) do not need or use O_2 as a nutrient. In fact, O_2 is a toxic substance for these organisms which either kills or inhibits their growth. Obligate anaerobic prokaryotes may live by fermentation (e.g., methanogenesis), anaerobic respiration or bacterial photosynthesis.
2. **Aerotolerant anaerobes** are bacteria with an exclusively anaerobic (fermentative) type of metabolism but they are insensitive to the presence of O_2. They live by fermentation alone whether or not O_2 is present in their environment.

Oxygen conditions for the growth of these microbes are summarized in Table 5.

5. **Osmotic Effects or Ionic Strength or Water Availability**

Microbes contain approximately 70–90% water and if placed in a solution with a higher solute concentration, they lose water which causes shrinkage of the cell (plasmolysis). Depending on the water activity, microorganisms are classified into three broad categories:

- Halophiles—Organisms growing optimally at the water activity of seawater (3–3.5% salt or NaCl concentration). Halophiles are subdivided into:
  - mild halophile: 1–6% NaCl (sea water is 3.5% salt)
  - moderate halophile: 6–15% NaCl
  - extreme halophile: 15–30% NaCl
- Osmophiles—Organisms living in high sugar environments
- Xerophiles—Organisms living in very dry environments.

Table 5 Categories of bacterial species based on their oxygen requirements

Category	Environment		Effect of oxygen
	Aerobic	Anaerobic	
Obligate aerobe	Growth	No growth	Required (utilized for aerobic respiration)
Microaerophile	Growth if O_2 level not too high	No growth	Required but at levels below 0.2 atm
Obligate anaerobe	No growth	Growth	Toxic
Facultative anaerobe	Growth	Growth	Not required for growth but utilized when available
Aerotolerant anaerobe	Growth	Growth	Not required and not utilized

6. **Moisture Content**

All microbes need water or moisture to grow. In the absence of moisture, bacteria form endospores which do not reproduce but allow them to survive until moisture and nutrients are available. Similarly, eukaryotes like protozoa form cysts or oocysts to survive in adverse environments (Tortora et al. 2010). Generally, bioprocesses like composting are limited by the availability of moisture and therefore, moisture or water needs to be added externally in such cases.

5 Microbes in Bioprocessing of Solid Waste

Several bioprocesses have been used for treating solid waste and include composting, vermicomposting, biogas or biofuel generation, and anaerobic digestion. Important microbes or microorganisms that are involved in the biological processing of solid waste include the following:

1. **Bacteria** Error! Bookmark not defined—These are ubiquitous, unicellular prokaryotic microorganisms. They can be autotrophs or heterotrophs. Bacteria are tremendously diverse in size, shape, physiology and nutritional requirements. They are always unicellular (single-celled) with shapes such as cocci (spherical), bacilli (rod-shaped), spirillum or vibrio (comma shaped) as shown in Fig. 7. A scanning electron micrograph (SEM) of *Escherichia coli*, a rod-shaped coliform and a common inhabitant of the gastro-intestinal tracts (guts) of warm-blooded animals is shown in Fig. 8. It is generally found in the feces of these animals and therefore, is present in soils, solid waste, water and wastewater and surface runoff.

   Several species of Bacteria and *Archaea* have been found in landfills and landfill leachate and many are known to be pathogenic (Gerba et al. 2011; Nimonkar et al. 2022).

   The most common genera that were found in landfill leachate from a landfill in Pune were *Bacillus* followed by *Clostridium* (Nimonkar et al. 2022). Of the Archaeal species, *Methanosarcina* and *Methanoculleus* were found in the same landfill leachate.
2. **Fungi**—These are multicellular, heterotrophic eukaryotic microorganisms. They grow well in low pH and moisture conditions. Fungi are extensively used in the production of organic acids, antibiotics and enzymes. The decomposition of solid waste in processes such as composting is due primarily to microbes like fungi and bacteria. The best examples of fungi are molds that grow on food and other surfaces in warm and humid environments.

   Fungi have been successfully used in solid waste management, including in the degradation of plastics. Table 6 is a summary of a few studies that used fungi for the treatment of solid wastes.

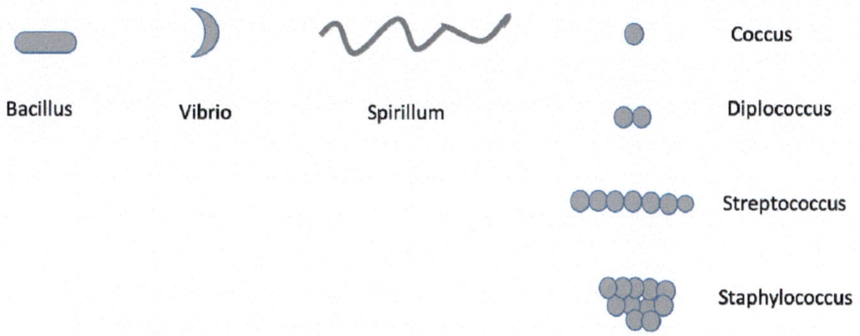

Fig. 7 Different shapes of bacteria

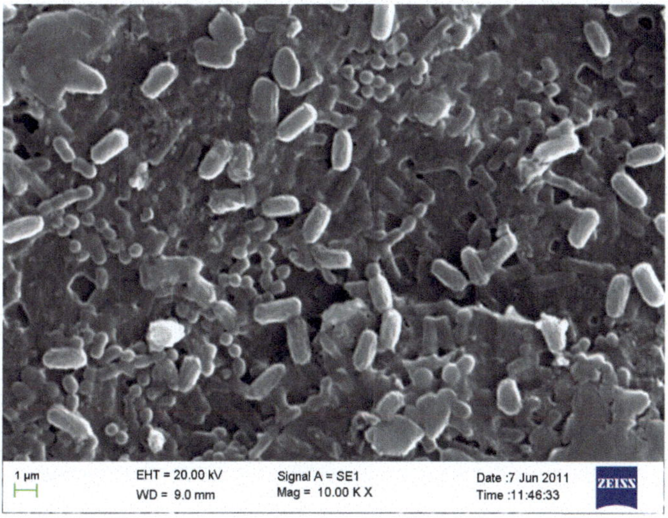

Fig. 8 SEM of *Escherichia coli*, a common bacterium (coliform)

6 Bacterial Growth and Its Quantification

Bacterial growth is defined by an increase in cell concentration measured in terms of numbers or cell mass. The growth rate is the change in cell concentration over a period of time, i.e., units of growth rate are cells/mL-time or g cells/L-time. In the case of bacteria, growth is typically quantified as an increase in cell concentration over a unit of time. Bacteria reproduce asexually by binary fission in which a bacterial cell divides and increases its cell number. Figure 9 depicts the steps involved in binary fission. A bacterial cell initially increases in size. Its circular chromosome (DNA strand) replicates itself within the large cell, the cell gets elongated and a septum

Table 6 Examples of fungi that can degrade different organic fractions of solid waste[4]

Type of waste	Fungi used for treatment	End product	References
Rubber industry and waste rubber	WRF4: *Resinicium bicolor*	Microbial devulcanization for recycling rubber	Asgher et al. (2008)
Polyester polyurethane	Endophytic Fungi: *Pestalotiopsis microspore*	100% clearance in 16 days leaving only fungal growth remains	Russell et al. (2011)
Cardboard (pretreatment for composting)	*Oligoporus placenta* and *Tremetes hirsuta*	Cardboard pretreated with fungi reduced cellulose, hemicellulose, and lignin. It is then composted in the normal way	Suthar and Kishore Singh (2022)
Food waste: avocado peel	*Fusarium sp.* and *Clonostachys sp.*	Compost	Becerra-Lucio et al. (2022)
Olive mill solid waste	WRF: *Anthracophyllum discolor* and *Stereum hirsutum*	>80% phenolic removal and 95% lignin degradation	Benavides et al. (2022)
Polyester polyurethane	*Embarria clematidis*	Up to 88% clearance of polyester polyurethane in 14 days	Khruengsai et al. (2022)

starts forming. Eventually, septum formation is complete resulting in two cells with a single DNA strand in each cell.

Bacterial growth curves in a batch culture

Bacterial growth is a result of both, change in cell size and replication of genetic material. Microorganisms can grow under a variety of physical, chemical and nutritional conditions. In a suitable nutrient medium, organisms extract nutrients from the medium and convert them into new biomass. Typically, nutrient media are formulated so that a specific nutrient, for example, carbon is the growth-limiting nutrient. In other words, all other nutrients are provided in excess of the organism's requirements and only one nutrient serves as the limiting nutrient. This nutrient or substrate is generally monitored in bacterial growth studies.

Growth Phases

A typical bacterial growth curve in batch culture includes the following phases and is shown in Fig. 10. Changes in biomass and substrate concentrations over time are shown in the figure.

1. Lag phase

[4] White root fungi.

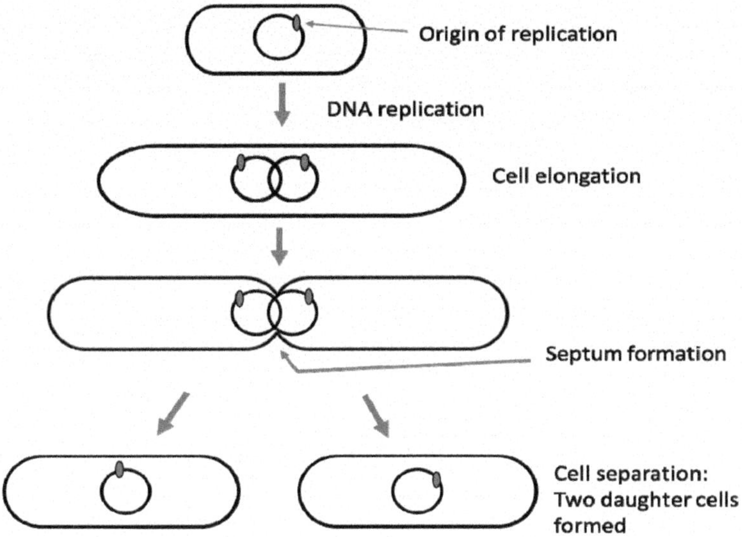

Fig. 9 Bacterial reproduction by binary fission

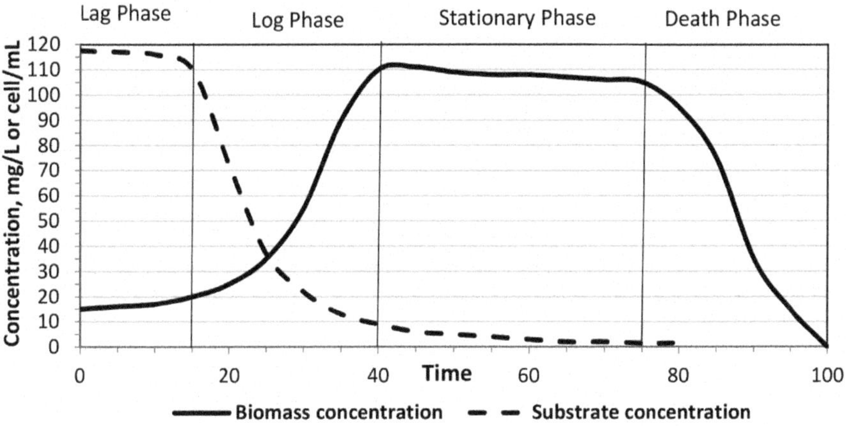

Fig. 10 Phases in a bacterial growth curve

2. Log phase or exponential phase
3. Stationary phase, and
4. Death phase.

1. **Lag phase**: The lag phase occurs immediately after inoculation and is a period of adaptation (or acclimation) of cells to a new environment.

   A long lag phase can occur when concentrations of nutrients and/or growth factors are low. For example, when *Enterobacter aerogenes* was grown in glucose and phosphate buffer medium, the lag phase increased with a decrease in the concentration of Mg^{2+} which is an activator of the enzyme phosphatase (Shuler and Kargi 2001).

2. **Log phase or Exponential phase**: The exponential growth phase is also known as the logarithmic growth phase. During this phase, cell mass and cell number density increase exponentially with time. Metabolic activity increases and the organisms begin replicating their DNA. Eventually, two copies of each cell are created and divided by binary fission. Growth occurs at a constant rate during this phase. The duration of this phase depends on nutrient availability.

   During the exponential phase, the net specific growth rate (μ) determined from either cell number-concentration or cell mass is constant. The exponential growth rate is first-order with respect to cell concentration, X.

$$\frac{dx}{dt} = \mu X, \text{ where } X = X_0 \text{ at } t = t_0 \quad (1)$$

   Integration of Eq. 1 yields,

$$\text{In}\, X/X_0 = \mu t, \text{ or } X = X_0 e^{\mu * t} \quad (2)$$

   where X and X_0 are cell concentrations at t and t = 0.

   The exponential growth is characterized by a straight line on a semi-logarithm plot of In (X/X_0) versus t.

   Doubling time is defined as the time required to double the microbial mass and can be determined using Eq. 2 by assuming that $X = 2X_0$. The result is Eq. 3 given below:

$$\tau_d = \text{In}\, 2/\mu = 0.693/\mu \quad (3)$$

   where τ_d is the doubling time of cell mass.

   Similarly, we can calculate a doubling time based on cell numbers and net specific rate of replication (μ_R). Thus, $\tau_d' = \text{In}\, 2/\mu_R$, where τ_d' is the doubling time based on the replication rate. In general, τ_d will equal τ_d', as the average cell composition and size do not change with time.

   Yield is defined as the amount of biomass produced per unit of the limiting nutrient (S) consumed. Growth yield is expressed as

$$Y_{X/S} = -\Delta X/\Delta S \quad (4)$$

3. **Stationary phase**: As the bacterial population continues to grow, nutrients in the growth medium are depleted by the microorganisms and the growth rate starts slowing down. At this stage, waste materials and toxic metabolites start accumulating in the medium, pH may decrease as acids accumulate, thereby creating an unfavourable environment for bacterial growth. Eventually, the growth rate equals the death rate of the cells and there is no net increase in cell concentration.

If an inoculum is taken from a stationary phase culture and is introduced into a fresh medium, the culture can grow easily again in the exponential phase.

During the stationary phase, one or more of the following phenomena may take place:

a. Biomass concentration may be constant, but the number of viable cells may decrease.
b. Cell lysis may occur and viable cell numbers may drop. A second growth phase may occur and cells may grow on lysis products of cells.
c. Cells may not grow but may produce secondary metabolites. Cellular regulation changes when concentrations of certain nutrients like Carbon, Nitrogen and Phosphate are low.

4. **Death or endogenous phase**: The depletion of nutrients and subsequent accumulation of metabolic waste products and other toxic materials in the media will result in a situation where the death rate is greater than the growth rate. During this phase, the bacterial cells start losing their ability to reproduce and often start utilizing their own biomass for survival; this phase is therefore termed endogenous phase. The number of dead cells may exceed the number of live cells. Some organisms can resist such conditions and can survive in the environment by producing endospores.

6.1 Monod Kinetics

In the 1930s and 1940s, Jacques Monod performed experiments on bacteria feeding on a single limiting nutrient in order to see if the logistic equation accurately described bacterial growth and published his results in 1949 (Monod 1949). The Monod model differs from the other models in the way that it introduces the concept of a growth controlling substrate or limiting nutrient. This model is similar to the Michaelis-Menten model used for describing the kinetics of enzyme reactions (published in 1913). When applied to bacterial cultures, bacterial growth kinetics can be described by the Monod equation:

$$\mu = \mu_{\max} \frac{S}{K_S + S} \tag{5}$$

$$\mu_{\max} = \mathrm{Yk} \tag{6}$$

Therefore,

$$\mu = \mathrm{YkS}/(\mathrm{K_s} + \mathrm{S}) \tag{7}$$

where S is the concentration of the limiting nutrient or substrate, $\mu_{\max}$ is the maximum specific growth rate, Y = yield and k = maximum substrate utilization rate per unit

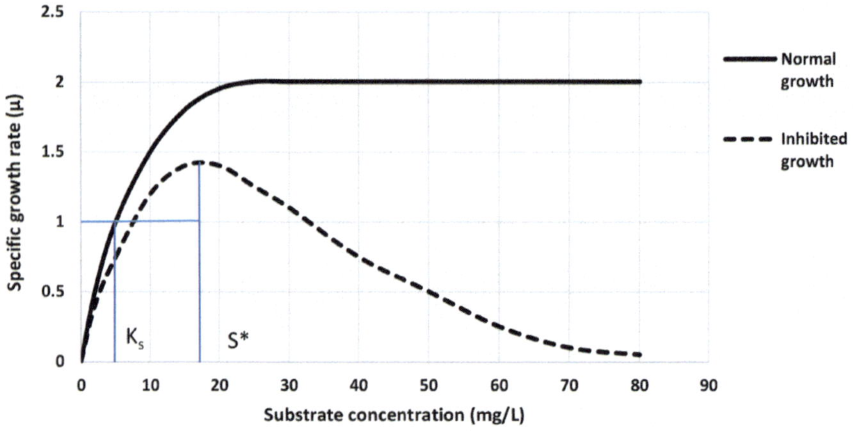

Fig. 11 Critical concentration of substrate (S*) under inhibited growth conditions

mass of cells. A graph corresponding to the above equation is shown in Fig. 11 and can be plotted by determining μ values for different S concentrations. The constant K_s is known as the saturation constant or half velocity constant and is equal to the concentration of the rate—limiting substrate when the specific rate of growth is equal to one half of the maximum. That is, $K_s = S$ when $\mu = 1/2\,\mu_{max}$. Furthermore, $\mu = \mu_{max}$ for $S >> K_s$ and $\mu = (\mu_{max}/K_s)S$ for $S << K_s$.

The Monod equation is analogous to Michaelis-Menten kinetics derived in 1913 for a single enzyme system where the amount of an enzyme or its catalytic activity is sufficiently low to make it reaction-rate limiting.

If endogenous decay is included in the growth equation:

$$\mu = \mu_m S/(K_s + S) - k_d \tag{8}$$

where k_d is the endogenous decay rate.

6.2 *Inhibition and Toxicity*

Toxic contaminants like pesticides, heavy metals and other synthetic organic compounds can inhibit the growth of microbes (LaGrega et al. 2001). As shown in Fig. 11, the presence of toxic or inhibitory substances will result in μ_{max} values that are lower than in the absence of these substances. Further, the onset of the death or endogenous phase will occur at an earlier time due to the presence of inhibitory substances.

The critical concentration of substrate does not follow Monod kinetics due to the presence of toxic substances and the specific growth rate is lower than that without toxic substances. In the figure above, S* is the concentration at which the maximum specific growth rate occurs.

7 Summary

Microbes are an essential part of many solid waste treatment processes. This chapter provides a simple and brief summary of the fundamentals of microbiology that are essential for understanding and controlling the behaviour of microorganisms in natural or engineered systems.

References

APHA, AWWA and WEF (2005). Standard Methods for Examination of Water and Wastewater. *In:* Standard Methods (12st ed.). APHA, AWWA and WEF, 2005. ISBN 9780875532356.

Asgher, M., Bhatti, H.N., Ashraf, M. and R.L. Legge (2008). Recent developments in biodegradation of industrial pollutants by white rot fungi and their enzyme system. *Biodegradation*, **19(6)**: 771–783. https://doi.org/10.1007/s10532-008-9185-3.

Becerra-Lucio, P.A., Labrín-Sotomayor, N.Y., Apolinar-Hernández, M.M., Becerra-Lucio, A.A., Sánchez, J.E. and Y.J. Peña-Ramírez (2022). Degradation activity of fungal communities on avocado peel (*Persea americana* Mill.) in a solid-state process: Mycobiota successions and trophic guild shifts. *Archives of Microbiology*, **204(1)**: 1–10. https://doi.org/10.1007/s00203-021-02600-3.

Benavides, V., Pinto-Ibieta, F., Serrano, A., Rubilar, O. and G. Ciudad (2022). Use of anthracophyllum discolor and stereum hirsutum as a suitable strategy for delignification and phenolic removal of olive mill solid waste. *Foods*, **11(11)**: 1587. https://doi.org/10.3390/foods11111587.

Binnig, G., Quate, C. and C. Gerber (1986). Atomic force microscope. *Physical Review Letters*, **56(9)**: 930–933. https://doi.org/10.1201/9781420075250.

Gerba, C.P., Tamimi, A.H., Pettigrew, C., Weisbrod, A.V. and V. Rajagopalan (2011). Sources of microbial pathogens in municipal solid waste landfills in the United States of America. *Waste Management and Research*, **29(8)**: 781–790. https://doi.org/10.1177/0734242X10397968.

Goel, S. (2019). *Water and Wastewater Engineering*. Cambridge University Press, Cambridge, Mass.

Khruengsai, S., Sripahco, T. and P. Pripdeevech (2022). Biodegradation of polyester polyurethane by *Embarria clematidis*. *Frontiers in Microbiology*, **13(June)**: 1–12. https://doi.org/10.3389/fmicb.2022.874842.

LaGrega, M., Buckingham, P. and J. Evans (2001). *Hazardous Waste Management*. McGraw Hill International.

Madigan, M., Martinko, J., Parker, J.B., 2003. The Biology of Microorganisms, 8th ed. Prentice-Hall Inc. Upper Saddle River, NJ.

Madigan, M.T., Martinko, J.M., Bender, K., Buckley, D. and D. Stahl (2015). *Brock Biology of Microorganisms* (14th ed.). Pearson Publishers. https://doi.org/10.1017/CBO9781107415324.004.

Metcalf and Eddy (2003). *Wastewater Engineering: Treatment, Disposal and Reuse*. Tata-McGraw Hill.

Monod, J. (1949). The growth of bacterial cultures. *Annual Reviews in Microbiology*, **3**: 371–394.

Nimonkar, Y.S., Kajale, S., Dake, M., Ranade, D.R., Yadav, K.K., Kumar, R. and O. Prakash (2022). A culture-based and culture-independent approach to the study of landfill leachate bacterial and archaeal communities. *Anaerobe*, **77**: 102626. https://doi.org/10.1016/j.anaerobe.2022.102626.

Russell, J.R., Huang, J., Anand, P., Kucera, K., Sandoval, A.G., Dantzler, K.W., Hickman, D.S., Jee, J., Kimovec, F.M., Koppstein, D. et al. (2011). Biodegradation of polyester polyurethane by endophytic fungi. *Applied and Environmental Microbiology*, **77(17)**: 6076–6084. https://doi.org/10.1128/AEM.00521-11.

Sawyer, C.N., McCarty, P.L. and G.F. Parkin (2003). *Chemistry for Environmental Engineering and Science* (5th ed.). McGraw Hill. https://4lfonsina.files.wordpress.com/2012/11/chemistry_for_environmental_engineering_and_science.pdf.

Shuler, M. and F. Kargi (2001). *Bioprocess Engineering*. Prentice Hall International.

Suthar, S. and N. Kishore Singh (2022). Fungal pretreatment facilitates the rapid and valuable composting of waste cardboard. *Bioresource Technology*, **344(PA)**: 126178. https://doi.org/10.1016/j.biortech.2021.126178.

Tortora, G., Funke, B. and C. Case (2010). Microbiology: An introduction. *In: E-book* (10th ed., Vol. 1). Benjamin Cummings. https://doi.org/10.1017/CBO9781107415324.004.

Whittaker, R.H. and L. Margulis (1978). Protist classification and the kingdoms of organisms. *BioSystems*, **10**: 3–18.

Woese, C. R., Kandler, O., & Wheelis, M. L. (1990). Towards a natural system of organisms: Proposal for the domains Archaea, Bacteria, and Eucarya. Proceedings of the National Academy of Sciences of the United States of America, 87(12), 4576–4579. https://doi.org/10.1073/pnas.87.12.4576

Chapter 11
Mechanical Biological Treatment of Municipal Solid Waste

Naseeba Parveen, Sukhdev Jaipal, and Sudha Goel

1 Introduction

Around the world, municipal solid waste (MSW) generation rates are rising. With rapid population growth and urbanisation, annual SW generation is expected to increase by 70% from 2.01 billion tons in 2016 to 3.40 billion tons in 2050 (Kaza et al. 2018). Compared to developed nations, residents in developing countries, especially the urban poor are severely impacted by unsustainably managed waste. In low-income countries, over 90% of SW is often disposed off in unregulated dumps or openly burned. These practices create serious health, safety, and environmental consequences. More sustainable methods are needed to address this issue.

The current regulations in India require the door-to-door collection of MSW and segregation of waste at source into 3 fractions of MSW: biodegradable, non-biodegradable, and recyclable materials. However, all urban local bodies (ULBs) have not been able to implement this and mixed MSW is common in most cities even at this time. Separation of the 3 fractions of MSW after collection is difficult to achieve manually and can be done using mechanical–biological treatment (MBT). MBT of unsorted MSW is an effective method to maximize the recycling and recovery of discarded materials and to facilitate the stabilisation of the biodegradable components. The general outline of MBTs is shown in Fig. 1. MBT is a generic term for

S. Goel (✉)
Civil Engineering Department and School of Environmental Science and Engineering, Indian Institute of Technology Kharagpur, Kharagpur 721 302, India
e-mail: sudhagiitkgp@gmail.com

S. Jaipal
Civil Engineering Department, Indian Institute of Technology Kharagpur, Kharagpur 721 302, India

N. Parveen
School of Environmental Science and Engineering, Indian Institute of Technology Kharagpur, Kharagpur 721 302, India

S. Goel (ed.), *Advances in Solid and Hazardous Waste Management*,
https://doi.org/10.1007/978-3-031-49144-3_11

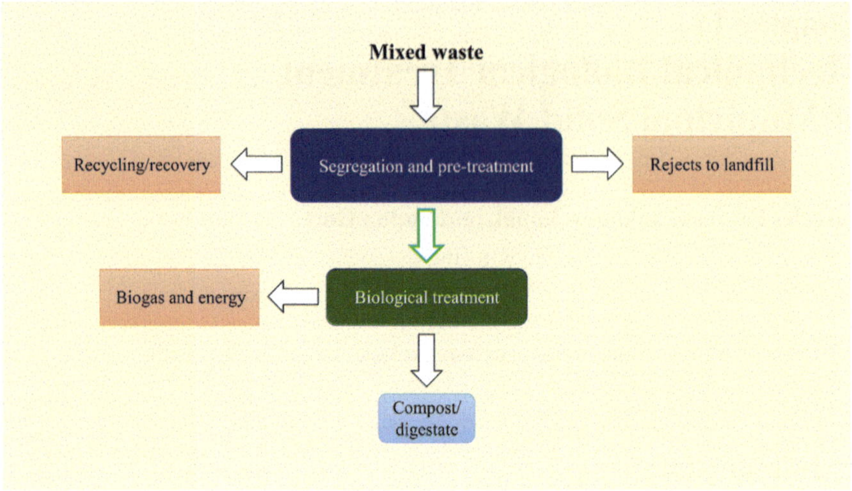

Fig. 1 General flowchart of mechanical biological treatment plants

the integration of mechanical and biological processes which are commonly used in SWM facilities such as materials recovery facilities, composting, and anaerobic digestion plants. The number of MBT plants, worldwide has grown in the last two decades as they can reduce the amount of waste that needs to be landfilled and allow the recovery and recycling of useful materials.

The objective of a Mechanical-Biological Treatment (MBT) plant may vary from place to place. In general, the main objective of MBT of municipal solid waste (MSW) is to minimise the amount of waste that is to be landfilled and to reuse and recycle the waste as much as possible (Bayard et al. 2010; Traven et al. 2018; Trois and Simelane 2010). MBT plants are planned and built to achieve other objectives as well:

- Pre-treatment of MSW prior to landfilling to reduce the volume and mass of MSW,
- Implementation of various biological and biochemical processes in order to reduce leachate and methane generation and odour emissions,
- Diversion of *non-biodegradable MSW* from landfills by using mechanical sorting facilities for material mining, recycling, or energy recovery as refused derived fuel (RDF), and
- Segregation of *biodegradable MSW* to create value added or useful by-products from waste such as compost or energy.

2 Scope for Mechanical Biological Treatment

The amount of MSW generated in urban areas is increasing exponentially due to the exponential increase in population and consumptive modern lifestyles. According to the World Bank, in 2016, the world's cities together generated 2.01 billion tons of MSW. Details of the generation and collection of MSW from selected cities are given in Table 1 (Kaza et al. 2018). The per capita generation of MSW in developed countries is much higher than that of developing countries. When most of the cities opted for door-to-door collection, some had curb-side pickup or centralized drop off points. Data regarding the number of trips required for transportation of MSW to the treatment or disposal site were not available for most of the cities. However, cities such as Kabul (Afghanistan), Gwalior, Ahmedabad, Vizag, Mumbai, Nagpur (all from India), Yokohama, and Osaka (Japan) take 2090621, 1133325, 935820, 862495, 831835, 721240, 698599, and 631225 trips per year to transport MSW from primary/secondary collection points to disposal/treatment facilities (Kaza et al. 2018). These numbers can be substantially reduced by building MBTs within the city which can act as collection as well as treatment facilities. The outputs of MBTs are value-added products for which customers can be found within the city itself.

Generally, MSW comes from households, commercial establishments, institutions and construction and demolition activities. Food waste, garden residues, paper, plastic, and inert materials are the major components of MSW. MSW contains both, organics and inorganics. The majority of the organic fractions are degradable (rapidly or slowly) while the smaller fractions tend to be non-biodegradable (e.g., plastic, rubber and leather). The rapidly degradable components include food waste and yard waste while paper, wood and textiles degrade slowly. Inorganic fractions include metals, glass, and inert materials (e.g., grit, soil, sand, and ash).

Globally, food and green waste contribute about half of the MSW (44%), followed by paper and cardboard (17%) and plastic (12%) as shown in Fig. 2. The term green waste includes yard waste and other vegetative wastes from markets. If the biodegradables are segregated, they can be further processed using various biological treatment units and can be converted to compost or digestate and biogas. Similarly, recyclables such as paper, cardboard, metals, and other valued components can be either reused or recycled. Apart from the wastes as shown in Fig. 2, there are special waste streams such as agricultural waste, construction and demolition (C&D) waste, e-waste, hazardous waste, industrial waste, and medical waste. Even though these categories are not considered MSW, they are managed under specific rules, but small fractions of these wastes are often found in MSW. By constructing MBTs, these special wastes can be segregated at the initial stage and their open dumping can be avoided.

As shown in Fig. 2b, most of the global MSW is either dumped openly (33%) or ends up in unspecified landfills, often without any prior treatment or segregation. This results in damage to the environment, contributes to the spread of various diseases, pollutes water, soil and air, and compromises the value of land in the surrounding area.

Table 1 Solid waste generation rate and collection mode in some of the major cities around the world (Kaza et al. 2018)

City and country	Population	Total MSW generation (tons/year)	Collection method	Transportation of MSW (trips/year)
Beijing, China	20384000	7903000	NA[a]	NA
Karachi, Pakistan	14910352	2719250	NA	254405
Mumbai, India	12478447	2497804.5	Door-to-door	831835
Sao Paulo, Brazil	11895893	4700000	Door-to-door	NA
Kinshasa, Congo	11587000	2555000	NA	NA
Moscow, Russia	11461000	5500000	Centralized drop-off point	NA
Delhi, India	11034555	3062350	Door-to-door	NA
Seoul, S Korea	10204000	3353985	Door-to-door	NA
Jakarta, Indonesia	9607787	2240735	Door-to-door	NA
Cairo, Egypt	9000000	5475000	Door-to-door	NA
Bengaluru, India	8443675	1277500	Door-to-door	20700
Tehran, Iran	8432000	2868170	Curbside collection	48834
London, UK	8173941	3560990	Door-to-door	NA
Rio De Janeiro, Brazil	6453682	3368499	Door-to-door	NA
Johannesburg, South Africa	4434827	1867000	NA	NA
Madrid, Spain	3165235	2652471	Other	NA
Toronto, Canada	2800000	942700	Curbside collection	NA
Dubai, UAE	2415000	2689808	Curbside	3.65E + 05
France, Paris	2254262	1092508	Door-to-door	NA
Minsk, Belarus	1915000	1135421	Centralized drop-off point	150200
Dhaka, Bangladesh	5700000	1424689	Door-to-door	239440
Seattle, United States	631350	720904	Curbside collection	NA

[a] Data not available

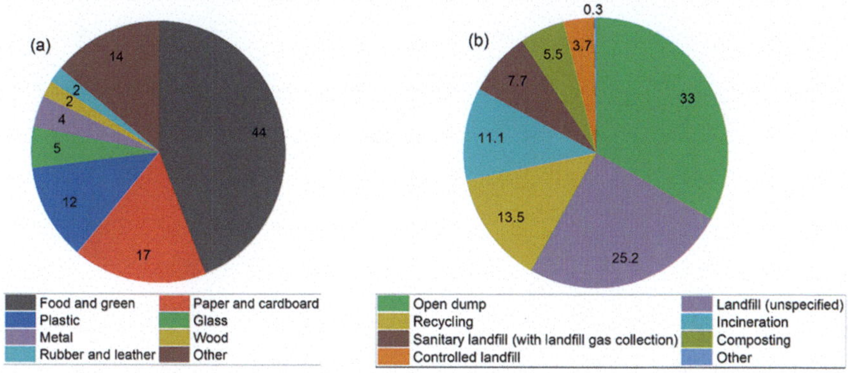

Fig. 2 Average global composition of MSW generated (**a**) and share of treatment and disposal methods for MSW (**b**). Based on data from Kaza et al. (2018)

3 MBT Processes and Materials Flow Analysis

Mechanical-Biological Treatment (MBT) plants have been setup in India and abroad and are operating successfully in many cases. Table 2 provides data regarding MBT plants in different cities in India and abroad and three plants: two from India and one from Germany are described in detail in this chapter.

Mechanical biological treatment processes may vary from plant to plant depending on the area's MSW generation and composition. Subsequently, the end products of the plant may also vary. However, some common processes like sorting, screening, and shredding are being adopted by several MBT plants. MBTs can operate with any combination of mechanical units and biological processes.

Some representative process units of an MBT from Bengaluru, India are shown in Fig. 3.

Mechanical units: The first process in MBT is the segregation of incoming waste for which several mechanical units are needed. Typically, these include a receiving floor, screens, shredders, and separation processes. From the mechanical units, materials for recycling or recovery are separated and the biodegradable materials are transferred to the biological units.

Biological units: Biodegradables can be treated aerobically or anaerobically in biological units.

Depending on the type of waste, an appropriate combination of units can be selected. A summary of MBTs and their processes is given in Table 2. The end products of most of the MBTs are compost, biogas, and RDF. To understand how MBTs operate, flowcharts for three MBT plants from Bengaluru and Goa in India and Hannover, Germany and their detailed working processes are described in this chapter.

Table 2 Mechanical biological treatment units and their processes from various parts of the world

References	Year	City and country	Plant capacity (ton/day)			Treatment flow			End products	Residue for landfill (%)
R. Bayard et al.	2009	Mende, France	25000	Sieve (70–450 mm)	Forced aeration (<450 mm)	Maturation	Fe-metal recovery			
Trois and Semilane	2010	Durban, South Africa		Segregation as wet and dry waste	Shredding		Mixing	Windrow composting	Compost	
Anurag Garg	2014	Delhi, India	7000	Sorting	Pre-trommel screen	Shredding	Air-classifier	Biological treatment (aerobic and anaerobic)	Compost or biogas, digestate, RDF	52
		Mumbai, India	7025	Sorting	Pre-trommel screen	Shredding	Air-classifier	Biological treatment (aerobic and anaerobic)	Compost or biogas, digestate, RDF	49.5
		Chennai, India	3500	Sorting	Pre-trommel screen	Shredding	Air-classifier	Biological treatment (aerobic and anaerobic)	Compost or biogas, digestate, RDF	46
		Kolkata, India	3000	Sorting	Pre-trommel screen	Shredding	Air-classifier	Biological treatment (aerobic and anaerobic)	Compost or biogas, digestate, RDF	43

(continued)

Table 2 (continued)

References	Year	City and country	Plant capacity (ton/day)			Treatment flow			End products	Residue for landfill (%)
		Hyderabad, India	2187	Sorting	Pre-trommel screen	Shredding	Air-classifier	Biological treatment (aerobic and anaerobic)	Compost or biogas, digestate, RDF	54
		Bengaluru, India	2000	Sorting	Pre-trommel screen			Biological treatment (aerobic and anaerobic)	Compost or biogas, digestate, RDF	3
		Ahmedabad, India	1683	Sorting	Pre-trommel screen	Shredding	Air-classifier	Biological treatment (aerobic and anaerobic)	Compost or biogas, digestate, RDF	54
		Pune, India	1175	Sorting	Pre-trommel screen			Biological treatment (aerobic and anaerobic)	Compost or biogas, digestate, RDF	11
Veilhaber et al.	2015	Hannover, Germany	540	Bag shredding	Screening	Air classification	Fe-metal recovery	Anaerobic digestion-aeration-maturation	Biogas	25
Trulli et al. (2018)	2017	Giovinazzo, Italy	320	Bag shredding	Magnetic separation	biological and thermal treatment			Electric energy, biogas	

(continued)

Table 2 (continued)

References	Year	City and country	Plant capacity (ton/day)			Treatment flow			End products	Residue for landfill (%)
Plavac et al. (2017)	2017	Celje, Slovenia	61	Bag shredding	Spreading	Sorting	Magnetic separation	Combustion	Compost, electric energy (MW), thermal energy (15 MW)	25
Mohanta and Goel	2021	Bengaluru, India	500	Screening			Fe-metal recovery	Windrow composting, vermicomposting	Compost	
				Sorting	Bag opening					

Fig. 3 Photographs from the Karnataka Compost Development Corporation MBT plant. **a** MSW receiving area, **b** trommel for waste screening, **c** composting windrows, and **d** final compost. Adopted from Mohanta and Goel (2021) with permission

Bengaluru, India

The MBT plant from Bengaluru receives mixed wastes from markets, households, and agricultural areas. The plant has a treatment capacity of 300 tons/day. The collected waste is tipped on a receiving floor; from there, it is passed through two different-sized trommel screens (200 and 100 mm size). The coarse fraction from the first trommel screen primarily contains used fabrics, banana fibers, inert materials, and plastic bottles, which are taken directly for landfilling without any treatment. The finer fraction from the 100 mm trommel is passed through a magnetic separator and ferrous metals are removed. The organics are moved to the biological unit for windrow or vermicomposting, after which another series of trommel screens follow (Fig. 4). The rejects of the 4 mm trommel screen are landfilled and the finer fraction from the same screen is taken for composting (Mohanta and Goel 2021).

Goa, India

The MBT plant situated in Goa, India is designed to treat MSW of various panchayats of North Goa (Tyagi et al. 2021). The capacity of the plant is 100 tons/day, and it receives 130 tons/day—173 tons/day of mostly dry and wet segregated waste. Around 18–25% of mixing in these was observed in the two streams. The received mixed waste is passed through a material recovery and segregation stage (MRSS)

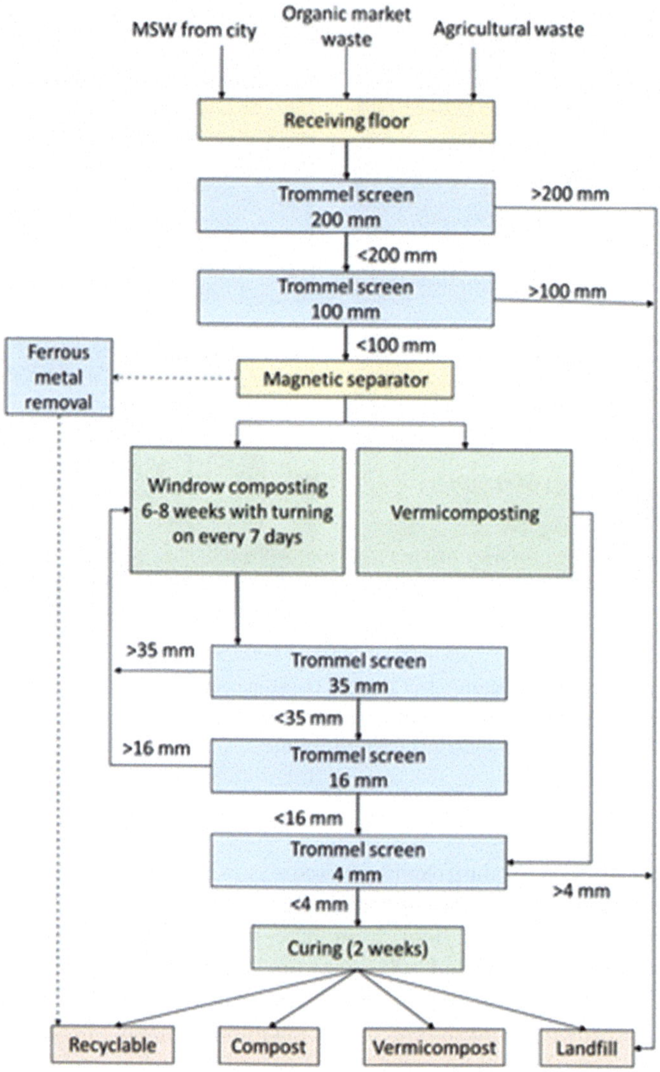

Fig. 4 Process diagram of MBT plant in Bengaluru (Mohanta and Goel 2021)

which includes a bag opening shredder followed by a roller disc screen (Fig. 5). If the incoming waste is completely dry rather than mixed, then this step is skipped. The waste is then divided into two lines from here. In the first line, the dry and large fractions are moved through a conveyance belt for manual segregation and in the second line, the underflow of the roller screen transfers the smaller fractions of the waste to an organic extraction unit (OREX). Line 1 comprises a manual sorting station where hazardous and e-wastes are first separated from recyclables.

Recyclables are then baled and sorted in containers prior to sending them off to the vendors. In the second line, OREX separates wet biodegradables or organics from dry inorganics through pressure exertion. The wet waste from OREX comes out in the form of organic pulped slurry which is then fed into an anaerobic digestor. Previously segregated wet waste can be directly added to this step skipping the former stages. The biogas generated in the digester is collected and cleaned after removing hydrogen sulfide and moisture and is used for the generation of electricity. A biogas yield of 5000 m^3/h with 65% CH_4 content was reported. The sludge from the digestor was composted in vessel composting drums (3 drums, six days retention). After separating the ferrous metals by a magnetic separator, the inorganic fraction of the OREX is shredded and screened through a double-deck flip flow screen followed by a wind sifter. The light fraction from the wind sifter is used for making the RDF and the coarse fraction is moved for landfilling. The RDF fraction is baled and stored for selling to cement companies.

Hannover, Germany

The mechanical treatment at the MBT of Hannover (capacity = 200,000 tons/year) includes the waste bunker, grabber, shredder, and series of screens and metal separators (Fig. 6). The waste unloaded in the shallow bunkers are placed in the shredder by grabber excavators. After removing the impurities, the first magnetic separator extracts the ferrous metals and the remaining waste is passed through a 60 mm screening drum.

The 60 mm screening drum separates the high calorific value coarse fraction such as wood, plastic, and paper from the finer fraction which is then passed through the second magnetic separator. About 55% of incoming waste at Hannover MBT is separated as a fine fraction at this stage and contains mostly organics suitable for fermentation. A second screening drum of 15 mm and an air classifier together prepare the fermentable portion of light weighted particles of <15 mm size.

The biological treatment unit of Hanover MBT includes three steps; (i) fermentation, (ii) aeration, and (iii) maturation (Vielhaber 2015). The light fractions from the air classifier are fed into an anaerobic digester where they undergo thorough dry fermentation for 20 days in the mesophilic temperature range. Biogas is extracted from the anaerobic chamber at this point and is purified and compressed before further use. In the aeration process, the digestate is thoroughly aerated for around 48 h, mainly to remove the ammonia content. After aeration, the sludge is matured in table windrows for another six weeks with a turning frequency of once a week. The resulting product is removed for landfilling and for final disposal. The mass balance for the Hannover SW plant shows that 46% of the total incoming waste is treated by MBT. After this, only a small fraction (24% of total waste) is landfilled. 43.5% of the plant's total waste that is not treated with MBT is composed of high calorific value coarse fraction that is incinerated and the remaining 1.5% is ferrous metals which are taken for recycling.

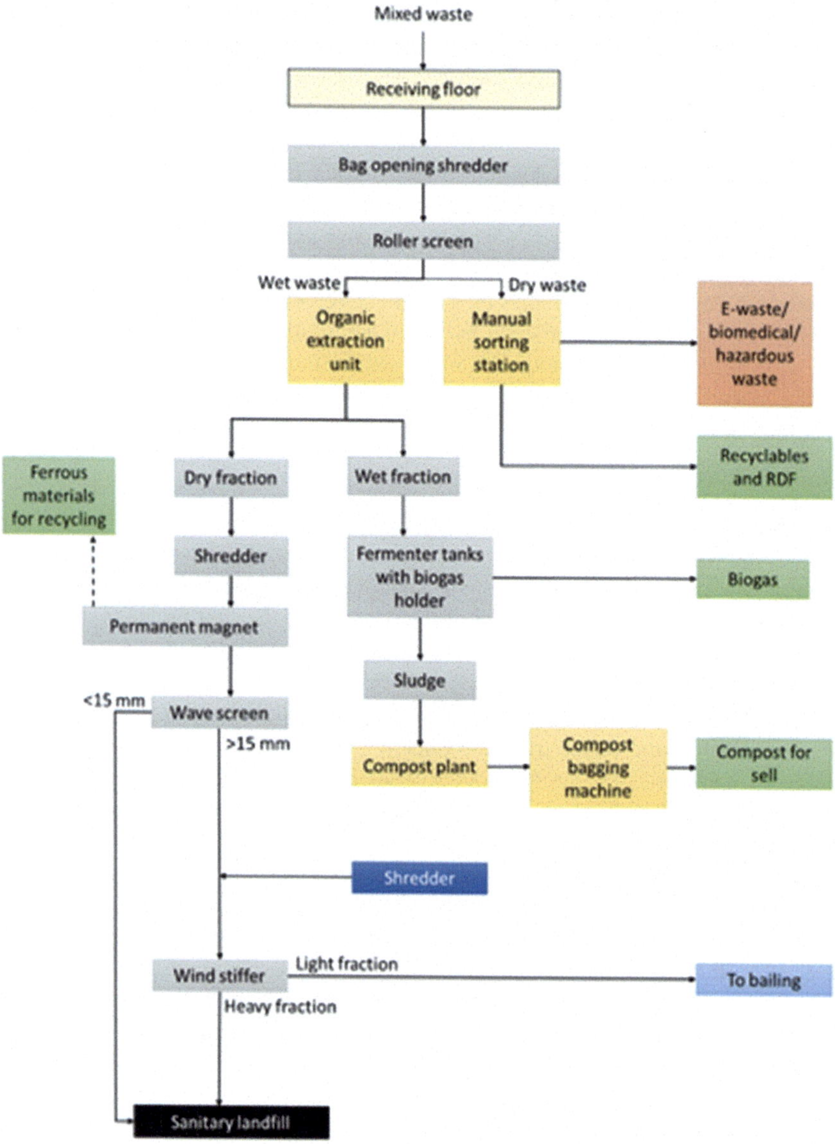

Fig. 5 Schematic of material process flow of MBT plant at Goa (Tyagi et al. 2021)

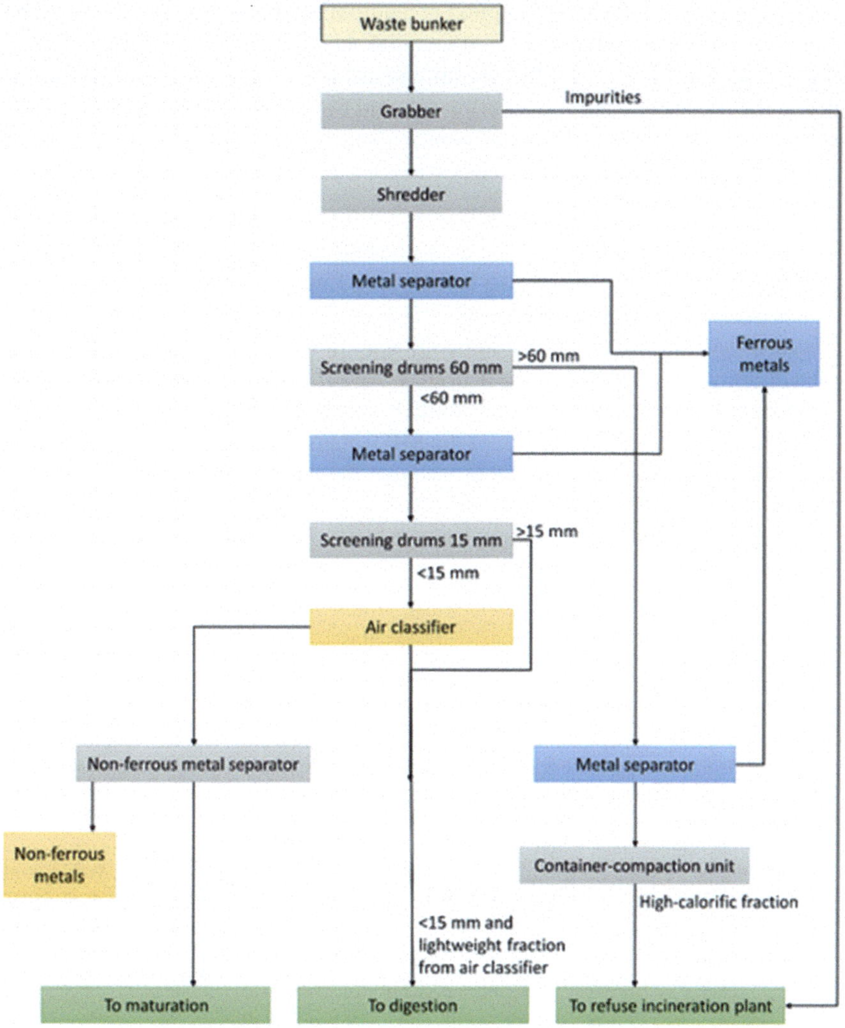

Fig. 6 Schematic of material process flow of MBT plant (mechanical units) at Hannover (Vielhaber 2015)

4 Conclusions and Suggestions for Improving the Efficiency of MBT Plants

Clearly, MBTs are suitable solutions for modern day MSW management of mixed waste and are potential methods for reducing the environmental impacts of open dumping. The materials flow analysis clearly indicates that MBT plants can be used to maximise material recycling and energy recovery. Detailed economic analysis (considering revenues from the outputs) is required before implementing any MBT design. The end users of the recovered material should be involved in the decision-making process to understand their requirements as well as to develop their confidence in the products. Appropriate efforts should be made to increase public awareness since sustainable waste management is extremely difficult to achieve without public participation. The following suggestions are made based on observations about the existing MBT plants.

1. Source separation should be done properly. Different types of waste should go in different types of streams of MBT plant. C&D waste should not be mixed with biodegradable and recyclable waste.
2. Data collected from cities should be reliable. In most developing countries, the availability and reliability of the data are generally poor. Not all waste collected is weighed. It is not clear how much waste is generated and delivered for disposal. This leads to inadequate design of the MBT plant.
3. Conversion of existing equipment, such as the use of the biological stage for waste drying can be done. It would be a recovery option for low-calorific fuel and additional separation of recyclable materials (such as pure-grade plastics).

References

Bayard, R., de Araújo Morais, J., Ducom, G., Achour, F., Rouez, M. and R. Gourdon (2010). Assessment of the effectiveness of an industrial unit of mechanical–biological treatment of municipal solid waste. *J. Hazard. Mater.* **175**: 23–32. https://doi.org/10.1016/j.jhazmat.2009.10.049.

Kaza, S., Yao, L., Bhada-Tata, P. and F. Van Woerden (2018). What a waste 2.0: A global snapshot of solid waste management to 2050? Urban Development, Washington DC, World Bank Publications. https://openknowledge.worldbank.org/handle/10986/30317. Accessed on 25/01/2022.

Mohanta, T. and S. Goel (2021). Mechanical-biological treatment of mixed municipal solid waste: Two plants in Bengaluru, India. *Asian J. Water Environ. Pollut.*, **18**: 9–17. https://doi.org/10.3233/AJW210014.

Plavac, B., Filipan, V., Sutlovic, I. and J. Sveticic (2017). Sustainable waste management with mechanical biological treatment and energy utilization/Odrzivo gospodarenje otpadom s mehanicko bioloskom obradom i energetskim iskoristavanjem. *Teh. Vjesn.—Tech. Gaz.*, **24**: 1283–1290.

Traven, L., Kegalj, I. and I. Šebelja (2018). Management of municipal solid waste in Croatia: Analysis of current practices with performance benchmarking against other European Union member states. *Waste Manag. Res.*, **36**: 663–669. https://doi.org/10.1177/0734242X18789058.

Trois, C. and O.T. Simelane (2010). Implementing separate waste collection and mechanical biological waste treatment in South Africa: A comparison with Austria and England. *Waste Manag.*, **30**: 1457–1463. https://doi.org/10.1016/j.wasman.2009.12.020.

Trulli, E., Ferronato, N., Torretta, V., Piscitelli, M., Masi, S. and I. Mancini (2018). Sustainable mechanical biological treatment of solid waste in urbanized areas with low recycling rates. *Waste Manag.*, **71**: 556–564. https://doi.org/10.1016/j.wasman.2017.10.018.

Tyagi, V.K., Kapoor, A., Arora, P., Banu, J.R., Das, S., Pipesh, S. and A.A. Kazmi (2021). Mechanical-biological treatment of municipal solid waste: Case study of 100TPD Goa plant, India. *J. Environ. Manage.*, **292**: 112741. https://doi.org/10.1016/j.jenvman.2021.112741.

Vielhaber, B. (2015). Mechanical-biological treatment plant in Hannover, Germany. Presented at the International Recycling and Recovery Congress IRRC Waste-to-Energy 2015, Waste Management, Vienna, p. 13. https://doi.org/10.13140/RG.2.1. 1513.8324.

Chapter 12
Current Status of Municipal Solid Waste Management in Kharagpur, India and Recommendations for Sustainable Management

Bishwatma Biswas, Naseeba Parveen, Mohammad Kalim Ahmad, and Sudha Goel

1 Introduction

Municipal Solid Waste Management (MSWM) is a major environmental issue at this time, mainly because of rapid population growth, economic growth and urbanisation. At the global level, the MSW generation rate is projected to produce 2590 million tons/year by 2030 (Kaza et al. 2018). Improper MSWM is a threat to public health and the environment (Goel 2017). Ideally, the MSWM system should be designed such that it safeguards the environment by reducing pollution and also helps to maintain clean and hygienic conditions in the cities. It also aims to minimise the cost of operation by resource recovery and energy generation (Planning Commission Report 2014).

Reliable data about total MSW generation rates and their compositions are needed for effective waste management and resource recovery. The development of an integrated MSWM plan requires knowledge of MSW generation rates, composition, collection, transportation and disposal as it helps to address the technical and financial requirements as well as the social and environmental impacts of MSWM. A brief overview of MSWM in India is provided here.

B. Biswas · N. Parveen · M. K. Ahmad
Environmental Engineering and Management, Department of Civil Engineering, Indian Institute of Technology Kharagpur, Kharagpur 721 302, India

S. Goel (✉)
Civil Engineering Department and School of Environmental Science and Engineering, Indian Institute of Technology Kharagpur, Kharagpur 721 302, India
e-mail: sudhagoel@civil.iitkgp.ac.in

S. Goel (ed.), *Advances in Solid and Hazardous Waste Management*,
https://doi.org/10.1007/978-3-031-49144-3_12

1.1 Solid Waste Management in India

Municipal solid waste management (MSWM) was a neglected issue in India until the Municipal Solid Waste (Management and Handling) Rules were passed in 2000. These Rules were modified to the Solid Waste Management (SWM) Rules, 2016. The SWM rules (2016) are now applicable to urban agglomerations, industrial townships, railways, airbases and special economic zones. Waste segregation at source was introduced to allow bioprocessing of waste and better resource recovery. Source segregation of residential waste is now mandatory; waste must be segregated into three fractions: biodegradable, non-biodegradable and domestic hazardous waste. India is shifting from an agricultural economy to an industry and services-based economy. World Bank data show that in 1960, agriculture, forestry and fishing together accounted for 41% of the country's GDP which decreased to 18.1% of the country's GDP in 2020 (World Bank 2021). Industrialisation and growth of the services sector have resulted in burgeoning urban populations and increasing MSW generation rates. As per the Census of India, urban areas in India accounted for 31.2% of the total population in 2011. India's urban population is expected to increase to 50% of the total population by 2053 based on an annual average exponential growth rate for urban areas of 2.85% (Goel et al. 2017). As the economy improves, it is expected that each person will generate more waste thereby contributing to an enormous increase in total MSW generation rates. The latest data from the Central Pollution Control Board (CPCB) show that more than 1.5×10^5 tons/day of MSW were generated in the country in 2018–2019 (CPCB 2019).

In 1989, the MSW collection efficiency in Indian cities ranged from 66 to 77%. In the last few years, it has improved from 82.5% in 2013 to almost 100% in 2018 based on CPCB annual reports as shown in Fig. 1. Before 2000, ULBs collected the bulk of the MSW generated and in general, the waste was dumped on open land without treatment. Improper MSWM is a threat to public health and the environment and implementation of the SWM Rules has resulted in major improvements in MSWM in the last two decades. With the passage of the SWM Rules, waste is now segregated at source in many Indian cities and properly designed treatment and disposal facilities are either operational or are being set up.

Globally, the trend is to reduce the use of landfills to the maximum extent possible and reduce, reuse and recycle waste as much as possible. The new goal is to send 'zero-waste' to landfills and countries such as Germany and Switzerland send less than 1% of the waste generated to landfills (Bufoni 2017). In the USA, the number of landfills decreased from 7924 in 1988 to 1269 in 2018 (USEPA 2019). In India, an initiative was taken to convert waste to energy (WtE) in the form of BioCNG, Biogas, or power thereby reducing the burden on landfills. The WtE programme aimed to set up 5000 biogas plants with a production capacity of 15 million tons of BioCNG by the year 2023–24 (MNRE 2022). The Swachh Bharat Mission is also aligned with the WtE program for converting MSW to power. The Ministry of New and Renewable Energy has approved the establishment of five WtE plants as of 31

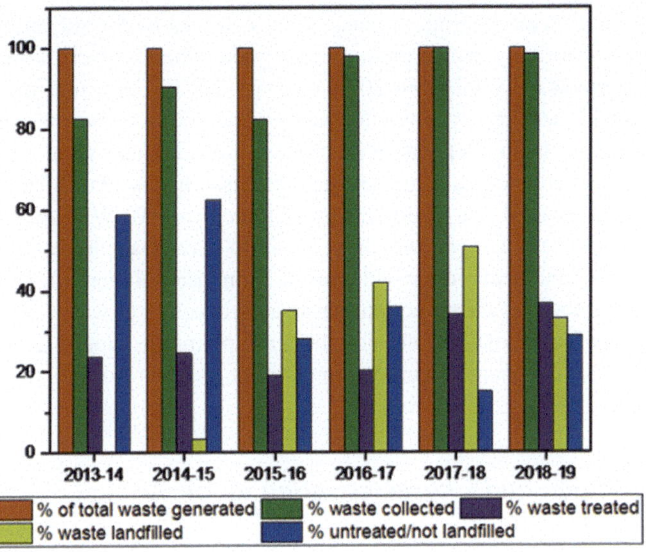

Fig. 1 Status of MSWM in India from 2013 to 2019 based on CPCB annual reports

January 2021 under the WtE program (MNRE 2022). These plants are expected to generate 74.7 MW of power from 6000 tons/day of MSW.

1.2 Environmental and Health Impacts of MSW

Unauthorised and non-scientific dumping of MSW on open lands can cause contamination of air, soil, and groundwater (Goel et al. 2017). Natural decay and combustion of MSW in open spaces results in the emissions of several gases, harmful volatile compounds and particulate matter in the air. Groundwater and soil below the dumpsite may be contaminated by the seepage of various biological and chemical products. One of the major factors of climate change and the increase in global temperature is the generation of greenhouse gases such as methane and carbon dioxide from organic waste decomposition. The SWM sector accounts for 11% of the total methane emissions and 3.2% of the total greenhouse gas emissions ("Global Methane Initiative" 2019; Herzog 2009). The burning of plastics present in MSW results in the release of many toxic gases such as furans, polychlorinated bisphenols, mercury and dioxins. Some persistent organic pollutants that are hydrophobic may sorb to plastic waste in the marine environment and bioaccumulate in the tissue of marine organisms resulting in ecotoxicity (Teuten et al. 2009). The presence of hazardous waste such as inflammable materials in MSW can also cause explosions and devastating fires.

The high organic content and high moisture content of MSW provide a favourable breeding ground for microbial pathogens and disease vectors such as mosquitoes, rodents, flies and fleas. Improper management can lead to chronic and contagious diseases in people living in nearby areas, rag pickers and workers associated with waste management (Thayyil et al. 2013; Annepu 2012). Falls, injuries, and respiratory, dermatological and eye infections were observed in workers associated with solid waste management. Unpleasant aesthetics and bad odours from waste disposal sites can cause psychological disorders and depression. The World Health Organisation (WHO) observed that improvement in the current MSWM systems in India can control or prevent diseases which will help to save financial resources that are currently spent on medical services (Planning Commission Report 2014). Many steps have been taken towards the reduction, reuse, and recycling of waste and more stringent waste disposal regulations are now being implemented due to concerns about human health, environmental degradation and loss of land resources.

1.3 Generation and Characterisation of Waste

A significant increase in the generation of MSW has been observed over time in India (both in urban and rural areas) and worldwide. However, due to the increase in urban populations and migration of rural populations to urban areas, the problem is more acute with urban MSW generation. MSW generation rates are usually quantified in terms of per capita generation rates (metric tons/capita-day) or total generation rates (metric tons/day). In 2016, 0.74 kg of waste was generated per capita per day globally. The generation of waste is correlated with urbanisation rates and income levels and it varies from 0.1 to 4.54 kg of waste per capita per day in different countries (Kaza et al. 2018). The National Environmental Engineering Research Institute (NEERI) and CPCB conducted a study in 2004–2005 on the generation of MSW in 59 cities (35 metro cities and 24 state capitals) in India and found that 39,031 tons/day of MSW were produced (CPCB 2013). Another study was conducted by the Central Pollution Control Board (CPCB) and Central Institute of Plastics Engineering and Technology (CIPET) during 2010–2011 and found that MSW generation increased by 29.6% in 6 years (CPCB 2013). Various sampling methodologies can be employed to determine the generation rates and composition of MSW. Estimation of MSW generation rates can be done using the load count method or weight and volume-based analysis (Goel et al. 2017). Sampling can be done through a door-to-door collection system or the waste collection and transportation trucks (Aguilar-Virgen et al. 2010). MSW composition depends on the average income levels or standard of living or Gross Domestic Product (GDP) of the region, the developmental status of the area, climate conditions, consumption patterns, environmental laws, recycling activities and housing structure. Due to the highly heterogeneous nature of MSW, gathering reliable data on the MSW composition is costly and time-consuming but it is necessary for better design and planning of MSWM systems (Goel et al. 2017).

The objective of this study was to evaluate the current MSW management system in Kharagpur City and find ways to improve it. A survey was conducted from

September 2019 to March 2020 to understand the current system and collect MSW samples from different parts of the city.

2 Description of Study Area

Kharagpur is situated in the Paschim (Western) Midnapore district of West Bengal. It is 134 km away from Kolkata (the capital of the state of West Bengal, India) and is linked via the National Highway (NH)-6. The coordinates of Kharagpur are Latitude 22°21′00″ North and Longitude 87°20′06″ East. Kharagpur was established as a major railway junction in India in 1898. The Kharagpur municipality was constituted in 1954 with five wards which increased to 25 wards in 1961. According to Census data, it had 35 wards and a population of 299,683 in 2011 (Census of India 2011). Based on Census data for the past 5 decades, it was estimated that the population in 2019 was 3,47,395 using the incremental increase method (Kharagpur Municipality, Data and personal communication with officials 2017). The total municipal area of Kharagpur is 111 km^2 with 13 markets, 22 hotels and 4 hospitals. Railway settlements along the railway line are not part of the municipality of Kharagpur.

3 Materials and Methods

3.1 MSW Sample Collection and Preparation

Waste samples were collected from four different community bins for five working days (excluding Sunday and Monday) during the period from October to December 2019. The four community bins (bins 1–4) are in Wards 11, 14, 22 and 28, respectively. Bin 1 is located near Biswakarma Mandir in Ward 11 which is adjacent to the railway track. Most residents in this area are from low-income groups. Bin 2 is for the Malancha Road community in Ward 14. This is one of the largest bins in the municipality area and is located on the road connecting to NH-6. This area comprises shopping complexes and commercial places. Wastes from six adjacent wards are deposited in this community bin before it is transported to disposal sites. Bin 3 is located near the bus stand at Kamal Nagar in Ward 22. Usually, waste from this bin is collected more frequently than the other bins to avoid unhygienic conditions. Many government offices are present in this area. The railway track is also adjacent to this area Bin 4 is located near the Kharagpur municipality office in Ward 28. This place has a few underprivileged human settlements such as Goyalapara, Jhapetapur, and Bichpara where basic living conditions are not adequate. In these areas, wood, coal, and cow dung cakes are used as fuel for cooking leading to the generation of ash in huge quantities. Further, some schools are also present in this ward leading to the generation of institutional waste.

The MSW community bins are large concrete structures. Every time before sample collection, the contents of the community bin were mixed. This was followed by taking 20–25 kg of mixed solid waste on a plastic sheet laid on the ground. Later, a representative sample was obtained by coning a pile of waste. The waste samples were then segregated manually depending on the physical components such as paper, plastic, textile, glass, metal, cardboard, gardening waste and miscellaneous waste. Each of these components was weighed using a portable electronic weighing machine and the weight of each fraction of the total solid waste was noted. All steps of the sampling method are shown in Fig. 2a. After the segregation of the separable physical components, the resulting material was a uniform blend of organic substances along with sand, dirt, mud and other inert material, which could not be separated manually. This mixture is termed a mixed residue. A total of 2 kg of the mixed residue was collected from each pile of the sample in polyethene bags. It was immediately taken to the laboratory for analysis of moisture content. After that, the dried sample was ground and sieved to prepare the samples required for proximate analysis (Fig. 2b).

3.2 Waste Characterisation

The collected samples were processed and evaluated for moisture content, volatile solids, fixed carbon, and CHNS elemental analysis. Moisture content was measured as per ASTM (E871-82); for that 25 g of each sample was taken in a crucible and measured in triplicate. It was then dried until a constant weight was observed in an oven at 105 °C for a period of 24 h. After cooling in a desiccator to room temperature, the weights of the samples were recorded separately. The percentage of weight lost in drying was calculated and noted as moisture content. Total solids are defined as the weight of the solid samples left after drying to a constant weight at a temperature of 103–105 °C. Fixed solids and volatile solids are the fraction of solids left and lost, respectively, after ignition of the sample in a muffle furnace at 550 ± 50 °C and these were determined as per the IS:10158-1982 method. A CHNS-O Elemental analyser (Euro Vector EA3000) was used to determine the elemental composition of MSW samples. For CHNS determination, dried MSW samples were weighed and poured into tin capsules before dropping them in the combustion chamber. A thermal conductivity detector (TCD) is attached to the instrument to detect the different elements.

4 Results and Discussion

A summary of the present MSW management system in Kharagpur is shown in Fig. 3. Waste is collected from various sources by the municipality. This waste is transported to various disposal sites in Kharagpur using tractors, trucks and tippers. In 2016, there were three open disposal sites at Nimpura, Ayma and Mirpur (Singhal

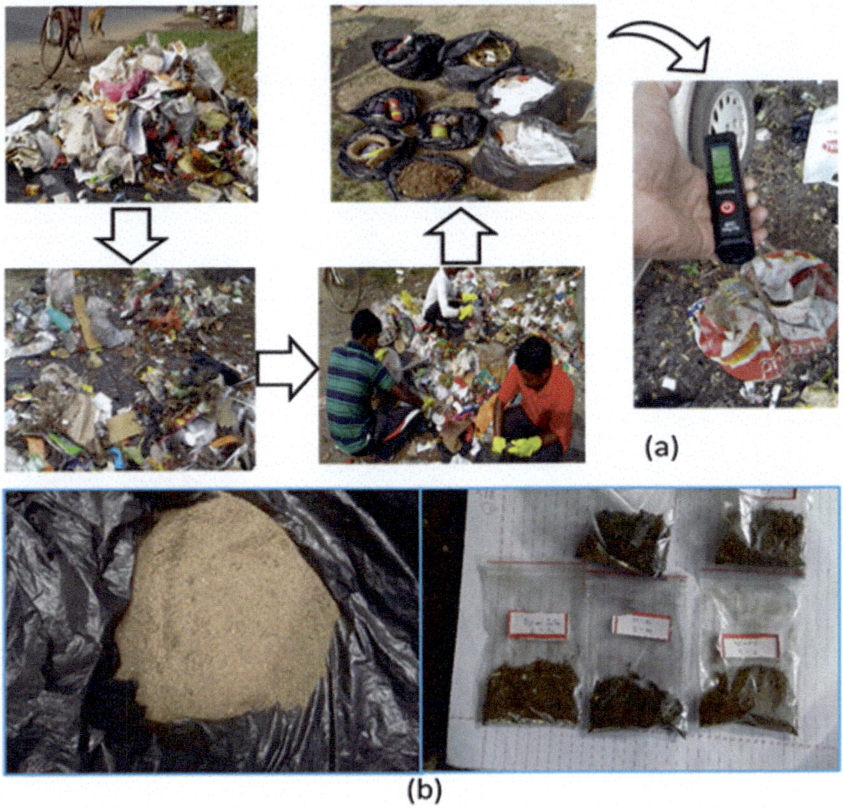

Fig. 2 **a** Steps involved in sample collection, segregation and weighing of the fractions using a portable electronic weighing machine. **b** Separated sand, dirt and other inert material from mixed residue and the final, fine ground sample

and Goel 2019; Kumar and Goel 2009). In 2019, another disposal site at Hiradihi was being used for open dumping of MSW.

4.1 Solid Waste Generation

Indian cities and towns produce 200–600 g/capita of MSW per day (Ministry of Environment and Forests 2016). Kharagpur Municipality officials estimate that the per capita MSW generation rate in Kharagpur city is 450 g/capita-day and the current population (2019) is 3,47,395. Therefore, the total MSW generated is estimated to be 156.33 tons/day. However, the total quantity of MSW collected by the municipality based on their records is 102.46 tons/day only, which means the collection efficiency is 65.5%. There are many areas in Kharagpur where waste is dumped on open land

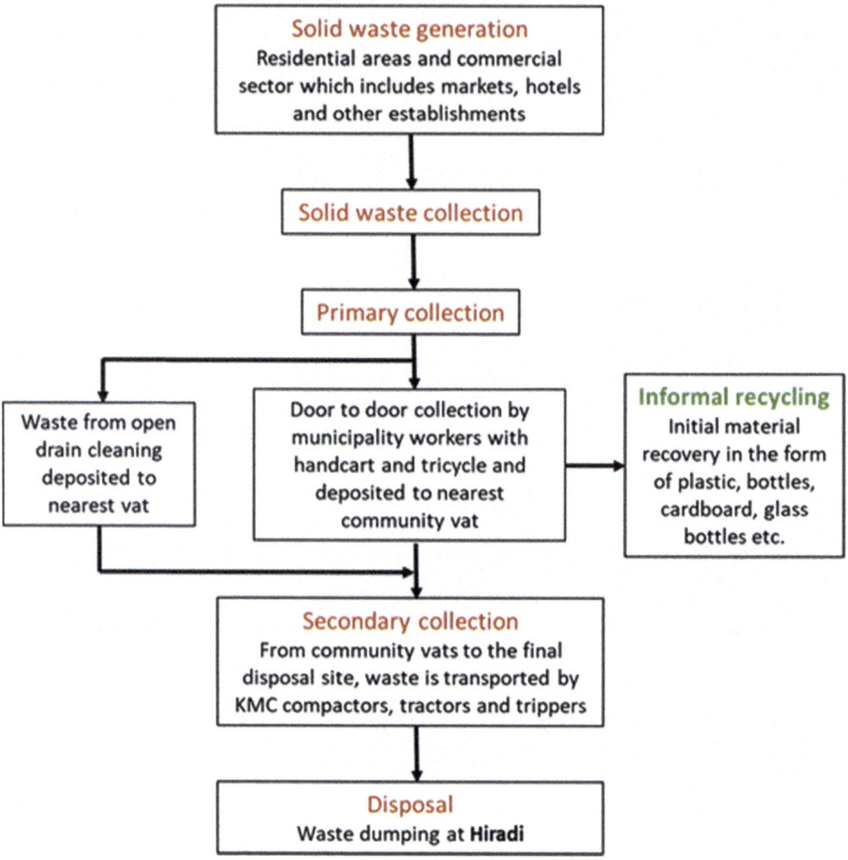

Fig. 3 Current MSWM system in Kharagpur City based on information provided by Kharagpur Municipality officials (2019)

and no collection is done. Sources of MSW in Kharagpur are similar to other Indian cities and include domestic (residential) waste, market, garden, commercial waste (hotels, businesses) and biomedical waste from hospitals and pathology laboratories. A breakup of the different sources of MSW is shown in Fig. 4.

In India, the bulk density of MSW varies around 400–500 kg/cubic meter as the waste is mainly street sweepings and inert material such as soil and sand (Ministry of Urban Development 2016). Compactor trucks and compactor bins are useful for wastes having low density such as packaging waste and plastic waste. These compactors are not cost-effective and provide little or no advantage for the transportation of high-density wastes with high food and garden waste content. Hence, compaction of waste is not necessary during transportation in Kharagpur.

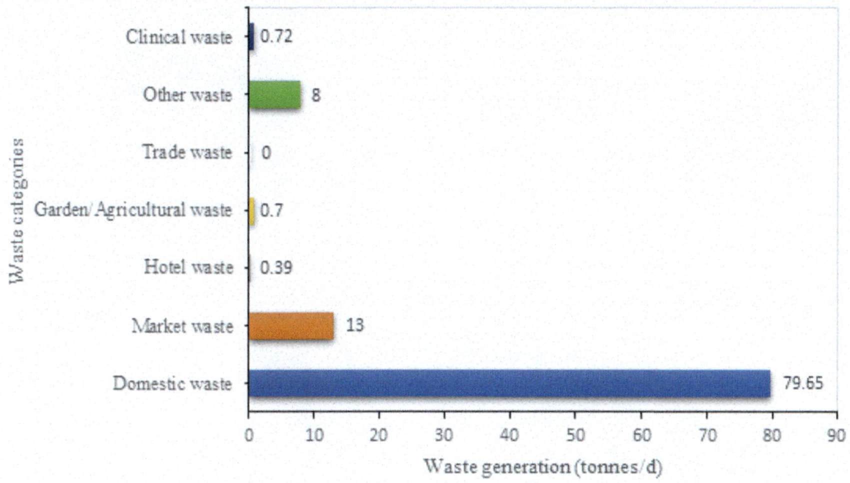

Fig. 4 Waste generated by different categories of sources in Kharagpur. *Source* Kharagpur Municipality officials (2017)

4.2 Present MSW Collection System

4.2.1 Primary Collection System of Solid Waste

Kharagpur municipality collects waste at the source of generation. This includes collecting waste from door-to-door from the residential areas which is also termed primary collection. The door-to-door collection is done by the waste collector, who blows a whistle to signal the household to deposit their waste in the collection vehicle. The collection is done in residential areas between 6:30 A.M. and 10:30 A.M. using tricycle carts and handcarts as shown in Fig. 5a.

Kharagpur municipality is facing a major problem in waste segregation as biodegradable and non-biodegradable wastes are collected together. A minor amount of resource recovery was found at the primary collection stage by the informal sector as shown in Fig. 5b.

4.2.2 Secondary Collection System of Solid Waste

After primary collection, the waste is deposited in a community bin, which is located in different parts of the ward. Roughly, 150 large and medium-size bins are placed in various parts of the city, mainly in market areas, major commercial buildings, and hotels. Such wastes are then transported to the dumping site. Every day compactors, trucks, tractors, and dumpers are used for secondary collection. During secondary collection, three workers collect the waste using different tools or a mini-wheel loader. A driver operates the vehicle/compactor/wheel loader. Figure 6 shows the solid waste

Fig. 5 **a** Tricycle cart and handcarts used for collecting waste in Kharagpur, West Bengal. **b** Collection of recyclable components of waste by the informal sector

collected from a community bin in ward 17 by municipality labourers, and waste collection by a wheel loader, and compactor truck in another ward. However, it has been observed that some amount of MSW remains uncollected and lies scattered in the city, making its way to the water bodies through drains and resulting in their pollution. Also, the municipality is not responsible for collecting and managing biomedical waste, radioactive waste, E-waste, and industrial waste. In these cases, the waste generator is legally responsible for the generation, collection, treatment and disposal of the waste.

4.3 MSW Characterisation

4.3.1 Physical Composition of MSW

Knowledge of the physical composition of MSW is very important as the amounts of recyclables present in the waste can be evaluated. After segregation, the composition of MSW was calculated in terms of the percentage of total (or wet) weight. Various physical components such as organic, plastic, paper and cardboard, textile, glass, metal, leather, rubber, and mixed residue were found and compared to the MSW of 2008 (Fig. 7). The average composition of MSW from each ward is shown in Fig. 7. Mixed residue contributed to 67–76% of the bulk waste. Despite manual segregation at the source, the mixed solid waste has around 10–14% plastic. Small amounts of

Fig. 6 **a** Community bins currently in use in Kharagpur, West Bengal. **b–d** Secondary collection of waste and transportation

other recyclable materials such as paper, textile, glass, metal and cardboard were also present in the bulk waste. Ragpickers pick up these recyclable wastes from community bins.

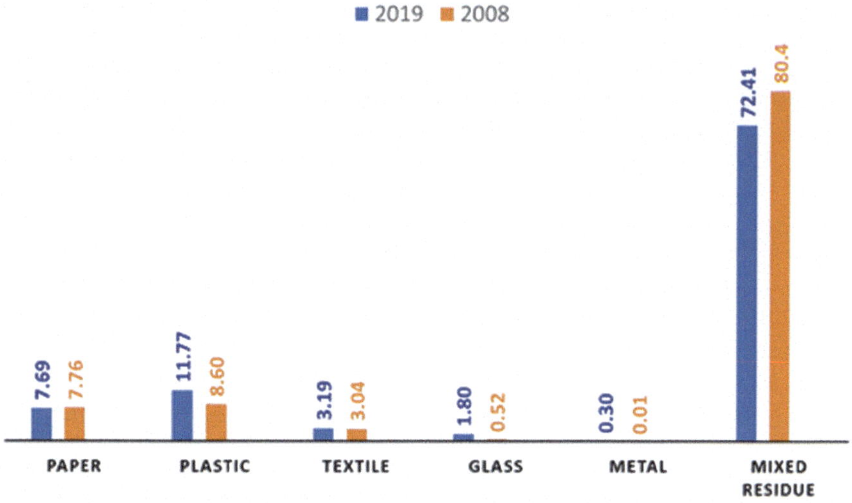

Fig. 7 The average composition of MSW (%) in 2008 and 2019

The composition of MSW from Kharagpur in this study was compared with that in 2008 (Kumar and Goel 2009). The percentages of plastic, metals, glass and textile waste were higher in 2019 compared to 2008. The plastic waste increased by 36.86%, paper waste generation was nearly the same, and the amount of mixed residue decreased by approximately 8%. Results of the MSW survey of Kharagpur from 2008 and 2019 showed high amounts of organic waste, especially food waste and plastic waste compared to other cities. These results indicate that composting or bio-methanation of MSW in Kharagpur are viable options if the segregation of waste into biodegradable and non-biodegradable fractions is carried out before or after collection.

4.3.2 Chemical Composition of MSW

A proximate analysis of the wastes collected from different wards in Kharagpur municipality was done and the average results are presented in Table 1. These data are necessary for determining the best available technology (BAT) for the treatment and disposal of MSW. The average moisture content of MSW was 29.43 $\pm$ 4.85% while the total solids were 70.57 $\pm$ 4.85%. The average values of fixed solids and volatile solids as a percentage of total solids were 71.6 $\pm$ 4.67% and 28.4 $\pm$ 4.67%, respectively. Thus, the average organic content (measured as volatile solids) of MSW was approximately 20% of the total or wet weight of MSW. In comparison with MSW from 2008, the organic content was 19.63 $\pm$ 9.53% of total solids or 11.5% of the total wet weight. The C/N ratio was found to be highest when the food market was held in the area. The amount of carbon in the samples was much higher than any other element and sulphur (an essential macronutrient) was absent which was evident from the CHNS-O elemental analysis. The values from this study can be compared to a similar study conducted in Kharagpur in 2008 (Kumar and Goel 2009). A significant decrease in moisture content was observed which led to an increase in organic content from 11.5% in 2008 to 20% in 2019.

4.4 Treatment and Disposal Facilities

As of 2019, there was no MSW treatment or disposal facility in Kharagpur. The municipality has a dumping site at Hiradihi. This disposal site is not operated or maintained scientifically. Mixed MSW is deposited in an open area without segregation. As there is no provision for creating a properly designed landfill facility, the present conditions lead to environmental contamination and degradation of natural resources around the existing disposal site. Treatment and disposal methods are usually selected based on the nature of the waste generated. Due to the high moisture and organic material content and low calorific value (800–1100 kcal/kg) of Indian MSW, incineration is not a viable treatment option. Based on the characterisation of the MSW generated in Kharagpur, the waste contains a large amount of mixed

Table 1 Results of the MSW characterisation of samples from different wards

Component	Average ± SD[a] 2019 (n = 20)	Average ± SD 2008 (n = 20)
Moisture content (%)	29.43 ± 4.85	42.05 ± 10.25
Total solid (%TS)	70.57 ± 4.85	58.36 ± 11.57
Volatile solid (% of TS)	28.4 ± 4.67	19.63 ± 9.53
Fixed solid (% of TS)	71.6 ± 4.67	80.35 ± 9.54
C/N Ratio	20.83 ± 6.96	
C (%)	15.73 ± 7.06	
H (%)	2.01 ± 0.87	
N (%)	0.89 ± 0.98	
S (%)	ND	

[a] SD: Standard Deviation

residue. The mixed residue is mainly composed of inert materials like soil, sand and other particulate matter and biodegradable waste. However, if source segregation of biodegradable waste can be achieved, the biodegradable fraction can be converted to compost or biogas as the biodegradable fractions of MSW are rich in organic content and moisture content.

Composting is the process of bacterial conversion of organic material present in MSW under aerobic conditions. The product after bacterial conversion is called compost which is useful as a soil conditioner in agriculture and horticulture. Biogas can be generated using anaerobic digestion (also called biomethanation) which is another suitable method for the treatment of Indian MSW. This method has been adopted in several cities such as Bangalore, Indore, Lucknow, Pune, Delhi, and several other cities. The final method of MSW disposal should be a sanitary landfill, where the waste is placed over a liner material and is covered with a layer of earth. This is the most common method of disposal used in developing countries. Further, as the current goal globally is 'Zero-waste to landfill', it is important to minimise the amount of waste that is to be disposed off in a landfill.

4.5 Comparison of MSW Composition in Kharagpur with Other Cities

Previous studies regarding MSW composition in Indian and foreign cities are summarised in Tables 2 and 3, respectively. Food waste contributed the highest to total MSW in all Indian cities except in Puducherry and Chennai. Yard waste (also known as green waste) was the largest component (38.4%) of the MSW in Puducherry, and the second largest in Chennai (32.35%). Inert materials (34.65%) contributed the highest to MSW in Chennai where food waste was only the third biggest component with only 8% in the total MSW (Palanichamy 2015). Similarly, in Puducherry, only 3.6% of the total waste was food waste which was a smaller component when compared with paper and cardboard (30%), plastic (10.4%), glass (5%), textiles (4.5%) and metals (4.1%) (Pattnaik and Reddy 2010). Puducherry Municipality has implemented three-bin segregation at the household level for biodegradables, recyclables, and other wastes. The food waste is dumped in the bin for biodegradables while the yard waste often remains unsegregated and is later taken by waste collectors to the dumpsite. The biodegradable content is collected twice a week and transported to composting/vermicomposting units. Therefore, the dump yards in Puducherry, from where Patnaik and Reddy (2010) in their study collected samples consists of less amounts of food waste. Besides food waste, inert materials contributed the most to total waste in most of the cities (Table 2). In Sangamner, Moradabad and Roorkee, the percentage of inert materials in total waste was more than that of food waste (Thitame et al. 2010; Siddiqui et al. 2012; Alam and Kulkarni 2016). The inert waste consists of materials such as sand, soil and dirt largely due to unpaved areas. A decrease in unpaved areas in Kharagpur during 2007–2019 because of new roads and buildings may be the reason for the decrease in the percentage of mixed residue. A comparatively small per cent of inert waste in Bengaluru also indicates a similar situation in Bengaluru, which is a well-developed metro city with paved roads and walkways (Speier et al. 2018). Metal was the smallest fraction of MSW, the least contributed component in the MSW. This is expected as most of the metal parts are either reused at the household level itself or are collected by rag pickers and 'recyclists' or 'kabaadiwalas' for recycling. This trend can be seen in other South Asian countries such as Malaysia, Pakistan, and Thailand as well (Table 3).

Compared to Indian cities, foreign cities had a higher proportion of paper/cardboard and plastic waste. A notable case was from South Korea, where food waste was much less than paper and cardboard waste, i.e., 8.1% and 35.1%, respectively (Ryu and Shin 2012). Muscat, Oman also had a similar composition, where food waste was only 12.31% while paper and cardboard contributed 19.25% of the total (Baawain et al. 2017). In both these cities, around 50% of the total MSW was composed of paper/cardboard and plastic together; 55.3% in South Korea and 48% in Muscat. Unlike the rest of the studies listed in Table 3, Oman and South Korea are developed countries in terms of infrastructure as well as economy. This finding is consistent with the World Bank Report that observed that high-income countries generate more dry waste (around 51% of total) such as plastic and paper compared to food waste unlike middle- and low-income countries where almost half of the total waste is wet or biodegradable waste (Kaza et al. 2018). The recyclable materials in low-income countries are often source-segregated and reused or sold as scrap and therefore the amount of paper/cardboard, plastic, glass, textiles, and metal fractions are less at the dump yards.

Table 2 Physical composition of MSW in various Indian cities

City	Stale	Population	Population year	Study period	No. of samples	Composition of the waste (%) Food yard waste	Glass	Metals	Paper and cardboard	Plastics	Rubber and Leather	Inert	Wood	Textile	Other	Total	References
Allahabad	Uttar Pradesh .	10,25,000	2003	Aug 2003–Apr 2004	22	45.3	0.73	2.54	4.69	2.86				2.22	41.66	100	Sharholy et al. (2007)
Kanpur	Uttar Pradesh	29,20,000	–	1988 1999	30	53.3 44.3	0.4	0.2 0.01	1.4 4	0.67 7.6	~	25.9 39.2	~	1.6 4.9	16.57	100 100	Zia and Devadas (2008)
Delhi	Delhi	1,37,80,000	2001	2002	–	38.6	1	0.2	5.6	6		34.7		–	13.9	100	Talyan et al. (2008)
Chennai	Tamil Nadu	43,43,645	2001	2000–2002	–	47.24	–	0.03	6.45	7.04	1.45	34.65		3.14	–	100	Jha et al. (2008)
Kharagpur	West Bengal	2,07,984	–	2007	20	–	0.52	0.004	7.76	8.6	–	–		3.04	80.1	100	Kumar and Goel (2009)
Puducherry	Puducherry'	9,73,829	2001	–	6	3.6 38.4	5	4.1	30	10.4	2.5	–	1.5	4.5	–	100	Pattnaik and Reddy (2010)
Sangamner	Maharashtra	61,958	2001	–2006	108	10 9.9	2	5	6.1	6	2.5	12.5	9.5	2.5	34	100	Thitame et al. (2010)
Moradabad	Uttar Pradesh	6,41,560	–	–	57	37.07	0.37	0.55	2.25	1.07	0.93	45.38	1.6	3.14	7.64	100	Siddiqui et al. (2012)
Jalandhar	Punjab	10,00,000	2011	2010	4	35	0.2	0.1	3.2	6.5	0.5	28	2	12.3	12.2	100	Sethi et al. (2013)
Varanasi	Uttar Pradesh	>1000000		2012–13	144	32	7	3	15	22	6	5	–	10	–	100	Srivastava et al. (2020)

(continued)

Table 2 (continued)

						Composition of the waste (%)											
City	Stale	Population	Population year	Study period	No. of samples	Food yard waste	Glass	Metals	Paper and cardboard	Plastics	Rubber and Leather	Inert	Wood	Textile	Other	Total	References
Chennai (2015)	Tamil Nadu	65,00,000				8 32.35		0.03	6.45	7.04	1.45	34.65	6.99	3.04		100	Palanichamy (2015)
Dhanbad	Jharkhand	2,90,000	2011	–	30	33	1.4	2.8		6.8	–	20.2	–	3.6	32.	100	Khan et al. (2016)
Roorkee	Uttarakhand	2,52,784	2011	–	8	17	20.9	1.98	3.66	9.6	3.47	–	19	–	1.9	100	Alam and Kulkami (2016)
Dehradun	Uttarakhand	5,83,679	2015	–	–	38		1	1	3	2	2	13	1	4	100	Joshi (2019)
Bengaluru	Karnataka	1,15,00,000	–	2015	6	51.9		1	0.2	7	5.5	–	3.6	0.7	2.7	100	Speier et al. (2018)
Varanasi	Uttar Pradesh	–	–	2013–15	12	46.13		2.56	0.92	9.77	10.2	–	–	–	2.77	100	Srivastava et al. (2020)
Kharagpur	West Bengal	3,47,395	2019	2019–20	20			1.80	0.30	7.69	11.76				3.19	100	Present study
																100	

Table 3 Physical composition of MSW in some cities around the globe

City	Country	Population	Study period	No. of samples	Food	Yard waste	Glass	Metals	Paper and Cardboard	Plastics	Rubber and leather	Inert	Wood	Textile	Other	Total	References
Chihuahua	Mexico	–	2016	560	48	–	5.6	2.4	16.1	11.9	–	–	–	–	16	100	Gomez et al. (2008)
South Korea	South Korea	–	–	–	13	–	–	–	33.9	21.4	–	7.3	7.9	8.1	8.4	100	Ryu and Shin (2012)
Bangkok	Thailand	>8 million	–	–	49.9		4.4	1.4	8.5	28.5	–	–	–	5.2	2.1	100	Laohalidanond et al. (2015)
Muscat	Oman	–	2014	4	12.31	6.08	3.89	3.28	19.25	28.75	–	2.15	1.49	14.37	8.43	100	Baawain et al. (2017)
Kyoto*	Japan	1.47 million	2012	268	39.8	2.3	0.9	2	32.2	9.7	0.7	0.5	0.8	–	11.1	100	Yamada et al. (2017)
Islamabad	Pakistan	3,56,603	2012	1260	59.95		2.79	1.68	7.26	7.49	–	–	–	1.81	19.02	100	Zia et al. (2017)
Beris Lalang	Malaysia	7,07,979	2018	–	34.64	1.77	0.85	1.02	17.59	27.17	6.28	–	1.38	3.68	5.62	100	Yusoff et al. (2018)
Astana	Kazakhstan	>1 million	2017–2018	–	47.2	–	6.2	2.7	12.5	15.4	3.4	–	–	–	12.6	100	Abylkhani et al. (2019)
Asturias	Spain	–	2013	–	38.1		5.6	3.6	20.6	10.8	–	–	–	10.9	10.4	100	Fernández-Nava et al. (2014)

5 Recommendations for Improved and Sustainable Management

Sustainable MSWM is possible only if the characteristics of MSW are known. A clear understanding of the percentage composition of MSW helps to create strategies to reduce, segregate, collect, reuse and recycle MSW. Sustainable and effective waste management can lead to energy generation in the form of biogas which in turn can be used to generate electricity and reduce greenhouse gas emissions.

5.1 Existing Scenario

Kharagpur municipal area produces 156 tons of solid waste per day, while the current solid waste collection system collects only 102 tons which indicates the need for improving the waste collection efficiency. Solid waste collected from each MSW generator is transferred to permanent community vats by manually-driven handcarts and tricycle carts. These existing old carts used for solid waste collection can be replaced by tractors and waste collected from door-to-door can be taken directly to the disposal site. At the secondary collection level, the number of community bins was found to be insufficient. Some wards do not have proper community bins and local people dump their waste on open land in their neighbourhood. The waste stored in the open vats is scattered by scavenging animals and birds, often creating an unhygienic and pathogenic environment in the vicinity. Though these vats are cleaned daily, they are not cleaned regularly. The following techniques are suggested based on the waste characteristics and current scenario for MSWM system for Kharagpur city.

5.2 Source Segregation and Transportation

The first step in any effective and holistic SWM system is the implementation of the 4R's Reduce, Reuse, Recycle and Recovery approach. Generally, plastics, glass, paper, textile and metal are considered to be recyclable materials, while wood and food waste are considered to be recyclable organic waste. These recyclable wastes should be separated from the MSW generation sources. This is regarded as the fundamental premise of MSW as it helps to improve recycling efficiency and reduces the cost of operation. A well-defined sorting method should be in place which will encourage the residents and businesses to separate their wastes at the source. The stationary containers currently in use need to be substituted with the hauled container systems. Hauled containers are favoured as they provide flexibility in choosing container locations. The covered bin also collects the leachate and prevents littering of wastes by birds or animals. The latest MSWM Rules, 2016, require that DTD collection should be implemented and no community vats or bins will be needed in the future. Small roadside bins may be needed for the collection of street litter. The implementation of source segregation should be ensured by regular monitoring and enforcement mechanism to verify residence compliance. Source segregation is a key component of a sustainable SWM system that ensures resource conservation and circularity.

5.3 *Organic Waste Management*

The mixed residue constitutes almost 71% (by weight) of the MSW generated, out of which 28.4% of the total dry weight of MSW is organic. Based on the properties of the organic content, the best treatment technology will be composting or bio-methanation for the organic fraction of MSW. As organic waste constitutes a large portion of MSW, composting can divert a large amount of waste from landfill. This diversion also helps to reduce many issues associated with it like leachate production and methane emission. The generated compost from the organic wastes can be used as a soil amendment which improves water retention, soil structures and fertility. Further, home-based or community composting can be introduced and encouraged. The incorporation of composting in SWM practice will help to effectively manage organic wastes, promote sustainable management, and reduce environmental impacts. It can be a key component of a sustainable SWM system.

Bio-methanation of the organic waste helps to convert organic waste into biogas and generate a nutrient-rich digestate. The generated biogas consists mainly of carbon dioxide and methane and methane is a flammable gas that can be used as a valuable source of renewable energy after it is separated from carbon dioxide. Biogas production also helps to reduce the use of fossil fuels and promote renewable energy production. The solid residue left after the bio-methanation process is known as digestate, which is a nutrient-rich material that can be used as an organic fertiliser. A report from EPA (2008) suggests that methane production from the organic fraction of MSW waste can be three-fold greater than from municipal wastewater through anaerobic digestion because of the greater organic content.

5.4 *Incineration and Landfill*

Incineration significantly reduces the volume of wastes that needs to be disposed off thereby minimising the space required for landfilling. Incineration helps to reduce the production of methane that is generated during the decomposition of waste in landfills (Fu et al. 2015). Modern age incinerators are equipped with advanced technologies which minimise the emissions of particulate matter and harmful gases. Incineration provides an effective method for the disposal of non-recyclable, medical and hazardous waste. It provides a secure and controlled means of destroying wastes which can pose a threat to the environment and public health if not disposed off properly (Wang et al. 2019). Though incineration offers such benefits, it is imperative to ensure that incineration facilities adhere to strict regulations and employ advanced pollution control strategies to ensure sustainable waste management.

The upgradation of open dumps and unsanitary dumpsites to sanitary landfills is imperative. Implementation of proper sanitary landfill minimises environmental impacts, and landfill gas collection system helps to utilise methane for energy generation. Sanitary landfills allow containment of the wastes, thereby preventing the risk of groundwater contamination, and odours associated with waste decomposition. Sanitary landfills are typically considered as the last resort for waste disposal; after all, efforts to reduce, reuse, and recycle waste have been maximised. Regular campaigns for increasing public awareness should be conducted to educate and sensitise the community regarding the benefits of sustainable SWM. Local governments and private sector companies can collaborate to establish sustainable waste management systems including composting, and waste-to-energy treatment plants. Life cycle analysis (LCA) can be done to identify the most suitable treatment technology for Kharagpur MSW.

6 Conclusions

The current study shows that Kharagpur municipal area produces 156 tons of solid waste per day, while the current solid waste collection system collects only 102 tons, indicating the need for improvement in waste collection efficiency. The solid waste fractions found in order decreasing quantities were: mixed residue (organic + inert) > plastics > papers > textiles > miscellaneous > glass > metals. Mismanagement of solid waste has remained one of the major problems of Kharagpur. The composition of MSW in 2019 was evaluated in the present study and compared with previously published literature. It was found that compared to Indian cities, foreign cities had a higher proportion of paper/cardboard and plastic wastes and lower production of inert waste. Low levels of awareness about the problems associated with poor waste management were less than 100% access to the DTD collection system and indiscriminate waste disposal in Kharagpur. This study suggests several key mitigation and treatment options including source segregation, composting/bio methanation, incineration, and finally, sanitary landfill to achieve sustainable SWM. The outcomes of this study are expected to be utilised for the development of a sustainable SWM policy, implementation, and treatment strategy.

References

Abylkhani, B., Aiymbetov, B., Yagofarova, A., Tokmurzin, D., Venetis, C., S. Poulopoulos et al. (2019). Seasonal characterisation of municipal solid waste from Astana city, Kazakhstan: Composition and thermal properties of combustible fraction. *Waste Management and Research*, **37**(**12**): 1271–1281. https://doi.org/10.1177/0734242x19875503.

Aguilar-Virgen, Q., Armijo-De Vega, C., Taboada-González, P.A. and S. Ojeda-Benítez (2010). Municipal solid waste generation and characterization in Ensenada, Mexico. The Open Waste Management Journal, **3**(**1**): 140–145. https://doi.org/10.2174/1875934301003010140.

Alam, T. and K. Kulkarni (2016). Municipal solid waste management and its energy potential in Roorkee City, Uttarakhand, India. Journal of The Institution of Engineers (India): Series A, **97**(**1**): 9–17. https://doi.org/10.1007/s40030-016-0146-2.

Annepu, R.K. (2012). Sustainable Solid Waste Management in India. Department of Earth and Environmental Engineering at Columbia University, pp. 1–189. Available from https://www.nswai.org/docs/Sustainable%20Solid%20Waste%20Management%20in%20India_Final.pdf. Accessed 17 June 2023.

Baawain, M., Al-Mamun, A., Omidvarborna, H. and W. Al-Amri (2017). Ultimate composition analysis of municipal solid waste in Muscat. Journal of Cleaner Production, **148**: 355–362. https://doi.org/10.1016/j.jclepro.2017.02.013.

Bufoni, A.L. (2017). The change from aerobic to anaerobic landfilling and increase of correspondent emissions and expenses. Modelling Trends in Solid and Hazardous Waste Management (1st ed., pp. 141–158). Singapore: Springer Nature.

Census of India (2011). Provisional Population Totals. New Delhi: Government of India.

CPCB (2013). Annual Report on Solid Waste Management (2013–14) Delhi, India. Available from https://cpcb.nic.in/uploads/MSW/MSW_AnnualReport_2013-14.pdf. Accessed 17 June 2023.

CPCB (2019). Annual Report on Solid Waste Management (2018–19), Delhi, India. Available from https://cpcb.nic.in/uploads/MSW/MSW_AnnualReport_2019-20.pdf. Accessed 17 June 2023.

EPA (2008). Anaerobic Digestion of Food Waste. Available from https://archive.epa.gov/region9/organics/web/pdf/ebmudfinalreport.pdf. Accessed 7 June 2023.

Fernández-Nava, Y., Del Río, J., Rodríguez-Iglesias, J., Castrillón, L. and E. Marañón (2014). Life cycle assessment of different municipal solid waste management options: A case study of Asturias (Spain). *Journal of Cleaner Production*, **81**: 178–189. https://doi.org/10.1016/j.jclepro.2014.06.008.

Fu, Z., Zhang, S., Li, X., Shao, J., Wang, K. and H. Chen (2015). MSW oxy-enriched incineration technology applied in China: Combustion temperature, flue gas loss and economic considerations. *Waste Management*, **38**(**1**): 149–156. https://doi.org/10.1016/j.wasman.2014.12.026.

Global Methane Initiative (2019). https://www.globalmethane.org/methane-emissions-data.aspx. Accessed 14 September 2021.

Goel, S. (2017). Solid and hazardous waste management: An introduction. Advances in Solid and Hazardous Waste Management (1st ed., pp. 1–370). Delhi: Springer and Capital Books. 1–27. https://doi.org/10.1007/978-3-319-57076.

Goel, S., Ranjan, V.P., Bardhan, B. and T. Hazra (2017). Forecasting solid waste generation rates. *In:* Modelling Trends in Solid and Hazardous Waste Management (pp. 35–64). Springer Singapore. https://doi.org/10.1007/978-981-10-2410-8_3.

Gomez, G., Meneses, M., Ballinas, L. and F. Castells (2008). Characterization of urban solid waste in Chihuahua, Mexico. Waste Management, **28**(**12**): 2465–2471. https://doi.org/10.1016/j.wasman.2007.10.023.

Herzog, T. (2009). World Greenhouse Gas Emissions in 2005. *World*, 2005–2009. Available from https://www.wri.org/research/world-greenhouse-gas-emissions-2005. Accessed 14 September 2022.

Jha, A.K., Sharma, C., Singh, N., Ramesh, R., Purvaja, R. and P.K. Gupta (2008). Greenhouse gas emissions from municipal solid waste management in Indian mega-cities: A case study of

their long-term effects (Schiavo et al. 2020; Thomas et al. 2015). The by-products of these plastics are also toxic to vegetation and possibly other organisms.

Compostable Plastics

Plastics that can be composted are made to break down in commercial composting facilities under controlled temperature, humidity, and microbial activity conditions (Greene 2007). They are expected to decompose into their natural constituent parts without leaving toxic residues. Plastics that adhere to specifications like ASTM D6400 or EN 13432 are considered certified compostable (Ciriminna and Pagliaro 2020).

Recyclable Plastics

Recyclable plastics are made to be gathered, processed, and turned into new goods; nevertheless, they are not directly connected to biodegradability. Recycling lessens the impact of plastics on the environment by preserving resources, preventing the generation of toxic by-products due to the decomposition of plastics, and generating less trash.

Bioplastics are plastics that are derived from renewable biomass sources, such as plants, and are intended to be environmentally-friendly alternatives to conventional petroleum-based plastics. They can be produced from various sources, including starch, corn, sugarcane, and cellulose. Bioplastics are designed to have a reduced carbon footprint compared to traditional plastics because their production relies on renewable resources and may require less energy. Further, the by-products of their decomposition are likely less toxic than those of conventional plastics.

The classification of plastics based on biodegradability highlights the need for careful consideration when selecting materials for various applications and waste management strategies, as well as the importance of proper labelling and disposal to minimise environmental harm.

3 Biodegradation of Plastics

Biodegradation of polymers occurs through various mechanisms: solubilisation, hydrolysis and enzyme-catalysed degradation (Shah et al. 2008).

Biotic degradation or biodegradation of plastic can occur under aerobic or anaerobic conditions.

i. *Aerobic biodegradation or respiration*: This is an important process in the natural attenuation of organic compounds where aerobic bacteria use oxygen as an electron acceptor and breakdown organic materials into final products of carbon dioxide (CO_2) and water (Nayak and Tiwari 2011)

$$C_{\text{plastic}} + O_2 \rightarrow CO_2 + H_2O + C_{\text{residual}} + \text{Biomass}$$

ii. *Anaerobic biodegradation*: Anaerobic biodegradation is also an important process in the natural attenuation of organic compounds where microorganisms break them down in the absence of oxygen. There are two possible pathways for anaerobic biodegradation: anaerobic respiration and fermentation. When anaerobic bacteria use nitrate, sulphate, Fe, Mn, and CO_2 as their terminal electron acceptors, the process is anaerobic respiration. Plastics can also be biodegraded by fermentation when the organic material (including plastic) serves as both, an electron donor and acceptor resulting in the production of smaller simple organic compounds (Nayak and Tiwari 2011) as shown below.

$$C_{plastic} + CH_4 + CO_2 + H_2O + C_{residual} + \text{Biomass}$$

3.1 Microbial Degradation and Their Mechanisms

Several studies have reported on the ability of microbes to degrade plastics. The following are a few studies on the microbial degradation of various grades of plastics.

3.1.1 Polyethylene Terephthalate

Polyethylene terephthalate (PET) is commonly used in the production of PET bottles, fibres, and foil. PET contains a high concentration of aromatic compounds, which prevent microbes from degrading the material.

Ideonella sakaiensis, a novel strain of bacteria, was found to be capable of degrading PET using two enzymes, PETase and MHEtase, yielding two products, terepthalic acid (TA) and ethylene glycol (Yoshida et al. 2021). Ribitsch et al. (2011) investigated the degradation of bis-(benzoyloxyethyl) terephthalate (3PET) using the enzyme p-nitrobenzylesterase produced by *Bacillus subtilis*. When 3PET was used as the sole carbon source, the bacterial strain produced 2-hydroxyethyl benzoate (HEB), mono-(2-hydroxyethyl) terephthalate (MHET), terephthalic acid (TA), and benzoic acid (BA). When cutinase enzymes from two strains of bacteria, *Thermobifida cellulosiytica* and *Thermobifida fusca*, were cloned into *E. coli* BL21-Gold (DE3), they demonstrated the ability to hydrolyse PET (Roth et al. 2014).

Cut190, a cutinase-like enzyme capable of hydrolysing the building blocks of PET, was synthesised using a bacterial strain of *Sacchromonospora viridis* (Miyakawa et al. 2015). Tcur1278, a polyester hydrolase secreted by *Thermomonospora curvata*, demonstrated positive hydrolytic activity against PET (Wei et al. 2014). Some filamentous fungi, such as *Fusarium oxysporum* and *Fusarium solani*, have also been shown to degrade PET fibres (Nimchua et al. 2007).

3.1.2 Polyethylene

Polyethylene (PE), one of the most widely used plastics, is classified into two types: low-density polyethylene (LDPE) and high-density polyethylene (HDPE). According to scientists, approximately 80% of total marine microplastics originate from PE.

Most *Bacillus spp.* and *Pseudomonas spp.* bacterial species are capable of degrading HDPE (Tsiota et al. 2018). A new strain of bacteria, *Achromobacterxylosoxidans*, was discovered by a group of researchers and has the potential to change the structure of HDPE (Kowalczyk et al. 2016). *Penicillium oxalicum* and *Penicillium chrysogenum*, two fungal strains, have demonstrated the ability to degrade polyethene (Ojha et al. 2017).

After a 28-day incubation period, Zhang et al. (2020) investigated the degradation of HDPE under the influence of a fungus *Aspergillus flavus* extracted from the guts of wax moths. In the laboratory, researchers were able to isolate a strain of *Klebsiella pneumoniae* that degraded HDPE (Awasthi et al. 2017). A cyanobacterial species known as *Nostoc carneum* was able to use the available carbon from LDPE as its primary source of energy, thereby degrading the polymer (Sarmah and Rout 2019).

3.1.3 Polyvinyl Chloride

Polyvinyl chloride (PVC) is a thermoplastic that can withstand harsh conditions and is widely used in household items as well as scientific and engineering activities. There have been very few successful studies of the biological degradation of PVC. However, some researchers have created a consortium of bacterial strains comprising *Pseudomonas otitidis, Bacillus cereus* and *Acanthopleurobacterpedis* isolated from Indian waste disposal sites (Anwar et al. 2016). Under ideal conditions, this consortium could degrade the structure of PVC. Some strains of the fungal species *Phaenerochaetechrysosporium, Lentinustigrinus, Aspergillus sydowii* and *Aspergillus niger* were able to grow in conditions where PVC was the only carbon source, indicating that they can degrade the polymer (Ali et al. 2014).

3.1.4 Polypropylene

Polypropylene (PP) is widely used due to its low cost, ease of processing and mechanical properties. It is used in the production of a variety of single-use disposables as well as long-term use durables. Its strength has been increased and its range of applications has been expanded by combining it with polyolefin rubber.

Researchers used the mesophilic bacterium *Strenotrophomonas panacihumi* to degrade PP. This bacterial strain was able to degrade both low and high molecular PP (Jeon and Kim 2016). Interesting studies have evaluated the rate of degradation of PP under the influence of bacterial strains isolated from mangrove sediments—*Bacillus sp.* and *Rhodococcus sp.* (Auta et al. 2018). In a study conducted by Mihreteab et al. (2019) fungal strains of *Yarrowialipolytica* were able to depolymerise polypropylene.

3.1.5 Polystyrene

Polystyrene (PS) is an aromatic polymer with high molecular weight derived from the monomer styrene. General-purpose polystyrene is a non-modified form that is physically clear and brittle (Ho et al. 2018). It has a low melting point and is a poor barrier to oxygen and water vapour (Ho et al. 2018).

Using a pure strain of *Pseudomonas aeruginosa*, researchers were able to successfully biodegrade modified PS (Shimpi et al. 2012). A pure strain of *Rhodococcus ruber* degraded a pure variety of PS in the form of flakes and powder (Mor and Sivan 2008). *Curvularia*, a fungal species, was exposed to a commonly used form of PS known as atactic PS. This fungal species was able to degrade PS and form an intricate network of hyphae that penetrated the PS surface (Motta et al. 2009).

Studies have shown microbes capable of degrading the most common types of polymers. To promote the degradation of plastics in the environment, it would be necessary to investigate the possibility of harvesting enzymes and other proteins from these microbes. The majority of studies were conducted in a laboratory setting, and applying them in the field necessitates additional knowledge.

3.2 Factors Affecting Biodegradation

The properties of plastics, both chemical and physical, directly influence their biodegradability and the mechanisms of their biodegradation.

3.2.1 Chemical and Physical Properties of Plastics

Physical properties of plastics that determine their biodegradability include surface conditions like surface area, hydrophilic and hydrophobic properties (Tokiwa et al. 2009), thickness of the material, and transmittance of light, especially UV light. The physical form of the materials such as films, pellets, powder or fibres (Arkatkar et al. 2009) also affects their degradation. Chemical properties of plastics that influence their biodegradability are first-order structures like chemical structure, molecular weight and molecular weight distribution, higher order structures like crystal structure, degree of crystallinity, and the presence of active functional groups (Arkatkar et al. 2009; Tokiwa et al. 2009).

Molecular Weight

Since microbial colonisation depends on surface characteristics that enable bacteria to establish a central locus from which they may spread their development, molecular weight plays a crucial role in the biodegradation of plastic. Biodegradation and molecular weight are inversely connected because an increase in molecular weight makes a polymer less biodegradable (Tokiwa et al. 2009).

Shape and Size of Plastic

The form and size of the polymers have a crucial effect on the biodegradation process. Large-surface-area polymers deteriorate more quickly than those with tiny surfaces (Tokiwa et al. 2009). Plastics can be in the form of pellets, fibers or films. Chinaglia et al. (2018) discussed how different plastics of different shapes and sizes tend to disintegrate. It was observed surface area and biodegradation rates were directly correlated, and that smaller polymers would rapidly biodegrade as compared to larger particles. Yang et al. (2005a, b) also discussed the dependence of the shape and size of plastics for biodegradation. It was noted that the shape of the plastics may have an impact on the availability of oxygen to the microorganisms. It was also observed that the shapes of the polymers influenced the early stage of biodegradation, while the biodegradation at the final stage was not impacted by the shapes.

Polymer Additives

Polymer additives—typically low molecular weight organic compounds—help microorganisms to establish colonies. The capacity of microorganisms to degrade plastics is affected by non-polymeric pollutants like dyes or filler materials (Ahmed et al. 2018). It is common knowledge that an increase in ligno-cellulosic filler will decrease the sample's thermal stability. The dispersion and interfacial adhesion between the ligno-cellulosic filler and the thermoplastic polymer system play a key role in the composite system's thermal stability (Czajka et al. 2022; Yang et al. 2005a, b). Similarly, metals can act as pro-oxidants in the production of polymers that are susceptible to thermo-oxidative breakdown from polyolefins (Mamin et al. 2023).

Polymer Crystallinity

The polymer chains are found in plastics, either in crystalline or amorphous form. Crystalline polymer chains have short-range order and align with one another to create ordered structures, in contrast to amorphous polymer chains, which have no order and their configurations are random (Zhang et al. 2021). Considering that bacteria tend to colonise polymer surfaces and use polymer compounds in amorphous regions of the polymer surface, polymer crystallinity can play a significant role.

Pre-treatment of Polymers

Pre-treating plastics using mechanical or chemical methods can sometimes increase their biodegradability by reducing their size and increasing their accessibility to microorganisms.

3.2.2 Environmental Factors

Depending on the environment in which the plastic is found, biodegradation rates can change. The activity of microorganisms and their capability to degrade plastic can be influenced by variables such as temperature, humidity, oxygen availability and pH levels.

Moisture

Moisture is crucial for the development and multiplication of microorganisms, and as a result, it has a significant impact on the biodegradation of polymers (Ahmed et al. 2018) A certain amount of moisture is required for the activation of microorganisms, and an increase in moisture content also increases bacterial ability for hydrolysis (Chamas et al. 2020).

Temperature

The softening temperature of a polymer has a considerable impact on its biodegradability. Higher melting point polymers are less susceptible to biodegradation. Enzymatic biodegradation is most effective when polymers with low melting points are used (Tokiwa et al.2009). For instance, purified lipase from *R. delemar* was capable of hydrolysing polyesters with low melting points (Calabia et al. 2010; Tokiwa et al. 2009). Different enzymes with distinctive active sites can biodegrade various kinds of polymer substrates. For instance, enzymes produced by the fungi *A. flavus* and *A. nigeras* degrade straight-chain polyesters produced from diacid monomers with 6 to 12C-atoms more quickly than straight-chain polyesters produced by other monomer sources (Kale et al. 2006). Petrochemical-derived polymers are resistant to environmental degradation because of their hydrophobicity and three-dimensional structure (Speight 2019; Yamada-Onodera et al. 2001). A study undertaken to understand the role of temperature in the biodegradation of plastics showed that temperature, in accordance with the Arrhenius equation, impacts the biodegradation rates (Pischedda et al. 2019).

pH

The rate of polymer hydrolysis can also be changed by adjusting the acidic or basic state. The development and reproduction of microorganisms, as well as the biodegradation of polymers, are all impacted by pH changes (Marin et al. 2013; Rowe et al. 2016).

Exposure to UV Radiation

The C–C bonds in the polymers can be broken by UV light (290–400 nm). Depending on their molecular structure, different grades of plastic exhibit varying rates of breakdown (Ranjan and Goel 2019, 2021). Plastics exposed to sunlight and weathering can become more susceptible to microbial attack.

Microbial Activity

Biodegradation also depends on the existence and activity of microorganisms that may degrade polymers. Various microbial communities with different capacities to break down plastic may exist in various situations. The length of time that plastics remain in the environment before being degraded can also affect the overall biodegradation rate. Microbes like bacteria and fungi can degrade plastics in mixed or pure cultures (Arkatkar et al. 2009; Shah et al. 2008). Factors affecting biodegradation rates of plastics include microbial species, moisture content, high temperature (50–65 °C), acidic conditions, nutrient availability as well as the nature and chemical composition of the waste (Kandakatla et al. 2013).

3.3 Biodegradation Tests

Biodegradation of plastics in the environment can be tested in several ways as listed below:

1. Field tests
2. *In silico* Simulation tests
3. Laboratory tests.

3.3.1 Field Tests

Biodegradation of plastic materials can be tested in field tests by burying samples in soil, placing them in a lake or river or performing a full-scale composting process. However, environmental conditions cannot be controlled and it is often difficult to be certain that loss of material weight is due to biodegradation only. Simulation tests are applied as an alternative to field tests (Shah et al. 2008).

3.3.2 In Silico Simulation Tests

Computational simulations and modelling play a crucial role in discovering and comprehending the biodegradation of plastics. These techniques can forecast and examine various aspects of plastic degradation. Computational techniques are used in *in silico* studies to offer explanations and pathways for plastic deterioration. The exploration of molecular interactions, the prediction of degradation routes, the identification of enzymes, and the use of machine learning to estimate the rate of plastic biodegradation can help to steer research.

The identification of the PETase and MHETase enzymes from *Ideonella sakaiensis*, which hydrolyse PET into its constituent monomers, opened the door to a viable pathway for PET biorecycling. Using molecular dynamics, several researchers have unravelled how enzymes bind to and degrade PET, paving the way for enzyme designs for the effective degradation of PET (Duru et al. 2021; Feng et al. 2021; Jerves et al. 2021; Joo et al. 2018). These findings imply that the employment of PET hydrolase enzyme in plastic hydrolysis will be beneficial in settings like the soil and the ocean, where the plastic deposits are mostly made of polyethylene terephthalate. A molecular dynamics study was also used to demonstrate that several plastic additives can promote plastic biodegradability (Zhang et al. 2022).

Computational docking has been used to find several plastic-degrading enzymes (Enyoh et al. 2022). The use of microorganisms and their enzymes represents the most environmentally friendly and effective decontamination method for cleaning our environment (Priya et al. 2022). Such methods suggest enzymes with high binding affinities to plastic molecules, directing further experimental work. Studies on molecular docking for enzyme degradation of polyethylene (Santacruz-Juárez et al. 2021), PET (Chen et al. 2022; Guo et al. 2022), PET nanoplastics oligomers (Duru et al. 2023a, b), polyester (Leitão and Enguita 2021) have proved successful.

Numerous factors affect how plastic degrades. Researchers can now concentrate on the major elements affecting deterioration with the help of Machine Learning algorithms that can find the most pertinent features and minimise data dimensionality (Duru et al. 2023a, b; Jiang et al. 2023; Zhu et al. 2022). Jiang et al. (2023) created a plastic enzymatic degradation (PED) framework based on machine learning to forecast the ability of an enzyme to break down certain plastics by discovering and identifying hidden patterns in protein sequences. Min et al. (2020) describe a data-driven approach to reveal degradation patterns of plastic debris by relating physical characteristics and molecular structures to biotic and abiotic degradation behaviour in seawater. The findings demonstrate a hierarchy of predictors to measure surface erosion as well as combinations of characteristics, such as glass transition temperature and hydrophobicity, to categorise ocean plastics into fast, medium and slow disintegration groups.

Therefore, these computational techniques collectively increase our understanding of plastic biodegradation by revealing molecular mechanisms, directing experimental work, forecasting degradation behaviour, and ultimately assisting in the creation of environmentally friendly approaches to managing plastic waste.

3.3.3 Laboratory Studies

The most useful biodegradation tests are laboratory tests. In these tests, defined culture media are used and inoculated with either a mixed microbial culture or pure cultures of microbes that may have been isolated or acclimated for a particular polymer. In laboratory tests, it is possible to conclusively determine degradation rates of plastics under 'optimal' but not natural conditions (Müller 2005). These tests have great value for systematic investigation and a deeper understanding of the mechanisms involved. Standard test methods have been used for a long time to evaluate biodegradation of different types of materials including plastics.

Standard Testing Methods

Visual Observations

Visual observations which include roughening of the surface, formation of holes or crakes, defragmentation or colour changes, and other images using Scanning Electron Microscopy (SEM) or Atomic Force Microscopy (AFM) can serve as indicative parameters of the biodegradation process. However, these parameters do not conclusively prove biodegradation in terms of metabolism.

Weight Loss Measurements

Weight loss measurements to determine the loss of polymer due to biodegradation is an easy and relatively inexpensive method. However, it provides no solid proof of degradation unless there is a simultaneous increase in microbial counts and appropriate controls are part of the experimental design.

Changes in Mechanical Properties and Molar Mass

As with visual observations, changes in material properties do not prove biodegradation of the material. Properties such as tensile strength is very sensitive to changes in the molar mass of polymer and thus can be considered as indicators of biodegradation processes.

Clear-zone Formation

Clear-zone test is a very simple qualitative method that is used to screen organisms that can degrade a certain polymer (Augusta et al. 1993). In this test, the polymer is dispersed as very fine particles within the synthetic agar medium, and this results in the agar having an opaque appearance. After inoculation with microbes, the formation of a clear halo around the colony indicates that the organisms are able to depolymerise the polymer which is the first step of biodegradation. However, this method was found to be more effective for screening dissolved compounds like antibiotics (Mohanta and Goel 2014) rather than solid materials like plastic films (Nithin 2013).

Agar Plate Tests

An agar plate test was developed for qualitative screening of biological growth in the presence of plastics (Nithin 2013). Standard petri plates were filled with nutrient agar and two grades of plastic film (cut to a diameter slightly smaller than the diameter of the petri dish) were placed on it. The plastic film was perforated with a needle as shown in the photos below. After a week of incubation, bacterial colonies were found in both grades of plastic proving that growth was possible on the surface of plastic (Fig. 1a, b). Bacterial growth in these experiments was attributed to growth on nutrient media in agar plates and the adhesion of colonies to plastic films.

Microbial Biomass Growth and Protein Content Assay

The evaluation of the protein content of microorganisms participating in the biodegradation of plastic can be used to track their development. An increase in microbial biomass may suggest an active breakdown of the plastic material. The amount of

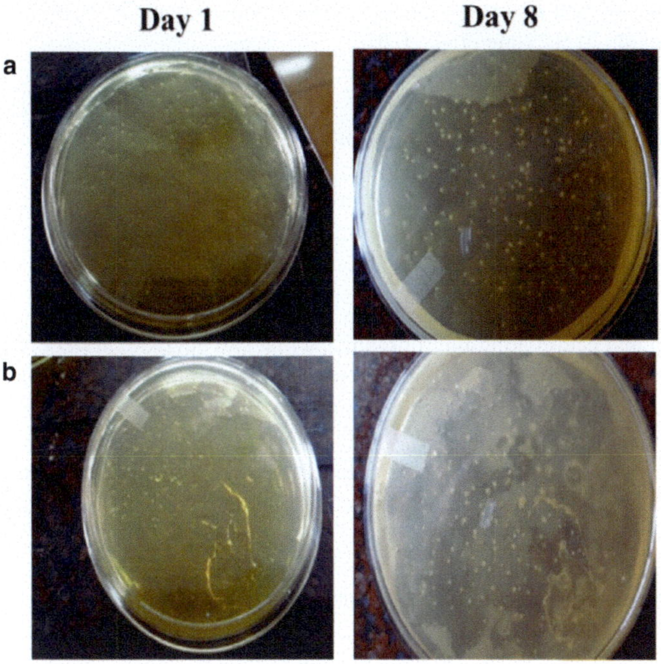

Fig. 1 **a** Grade 1 and **b** Grade 2. Agar plates show biological growth on plastic after incubation for 8 days

protein in microbial cultures can be determined using methods like the Bradford test, Lowry assay, or BCA assay (Giacomucci et al. 2019; Taniguchi et al. 2019).

Metabolic Activity Assays

In the breakdown of plastic, enzymes generated by microbes are crucial. One can gauge the activity of certain enzymes involved in the degradation of plastic, such as lipases, proteases, or esterases. Assays of enzyme activity can shed light on how microbes degrade materials (Ali et al. 2021; Atanasova et al. 2021; Syranidou et al. 2019).

Thermogravimetric Analysis (TGA)

Plastics can lose weight as a result of temperature fluctuations, and TGA can measure this weight loss. It offers details on the material's thermal stability and may be used to estimate the level of biodegradation (Khan et al. 2023; Yue et al. 2021).

Gel Permeation Chromatography (GPC) or Size-Exclusion Chromatography (SEC)

To examine how the molecular weight distribution of polymers has changed over time, utilise GPC or SEC. As the polymer chains disintegrate during biodegradation, the molecular weight frequently decreases (Giacomucci et al. 2019; Syranidou et al. 2019; Xie et al. 2021; Yang et al. 2020).

3.3.4 Enhancement of Biodegradation

Efforts were made to enhance the rate of biodegradation of these recalcitrant polymers by modifying the polymer or initiating the degradation process by generating free radicals, etc. (Arkatkar et al. 2009). Methods for enhancing biodegradation of plastics include (Bhardwaj et al. 2013).

- Isolating and acclimating microorganisms that can degrade these polymers.
- Improve the attachment of organisms to the recalcitrant polymer surface.
- Pre-treat the polymeric materials using thermal, UV, microwave, high-energy radiation and chemicals.
- Mix polymers or plastics with natural biodegradable polymers including cellulose or starch or with biodegradable synthetic polymers like polylactic acid (PLA) or Polycaprolactum (PCL).
- Mix with pro-oxidants.
- Use genetically-modified microbes for biodegrading these polymers.

4 Summary

The biodegradation of plastics is a multifaceted process influenced by a range of factors. Plastics have transformed many businesses and aspects of daily life, yet their enduring presence in the environment has led to serious ecological issues. Potential mitigation of the effects of plastic waste on the environment is provided via biodegradation. Nevertheless, this procedure varies depending on the kind of plastic and the surrounding environment. The kind of polymer, microbial activity, environmental elements like humidity and temperature, and the presence of additives all have a significant impact on how quickly and how much plastic degrades. Plastic-degrading microorganisms gradually convert these synthetic chemicals into simpler substances, potentially lessening their long-term impact.

Researchers are investigating numerous ways such as genetic modification of microorganisms, enzymatic treatments, and developing more biodegradable plastic substitutes, in an effort to harness and expedite plastic biodegradation. A comprehensive plan to combat plastic pollution must emphasise proper waste management, recycling, and reducing plastic usage. With the advent of computational modelling, it is essential to use *in silico* technologies to understand plastic biodegradation. They help in the design of biodegradable polymers and speed up research by enabling simulations of degradation processes, enzyme interactions, and toxicity predictions. These computational tools enhance the experimental design, combine data, and provide insights into the mechanisms of degradation, eventually assisting in the development of sustainable approaches to the management of plastic waste.

This chapter throws light on the different types of microorganisms that can initiate the biodegradation of different types of plastics. The emphasis is on the enzymes secreted by these microorganisms and the mechanisms followed during the degradation. Biological processes are slower than physico-chemical processes. However, further studies can help us understand how to enhance the degradation process. Interdisciplinary cooperation and cutting-edge research will be essential as we continue to understand the complexity of plastic biodegradation in order to provide sustainable solutions that protect our environment and future generations from the harmful consequences of plastic waste.

References

Abdelmoez, W., Dahab, I., Ragab, E.M., Abdelsalam, O.A. and A. Mustafa (2021). Bio- and oxo-degradable plastics: Insights on facts and challenges. Polymers for Advanced Technologies, 32(5): 1981–1996. https://doi.org/10.1002/PAT.5253.

Ahmed, T., Shahid, M., Azeem, F., Rasul, I., Shah, A.A., Noman, M., Hameed, A., Manzoor, N., Manzoor, I. and S. Muhammad (2018). Biodegradation of plastics: Current scenario and future prospects for environmental safety. *Environmental Science and Pollution Research*, **25**: 7287–7298. https://doi.org/10.1007/s11356-018-1234-9.

Ali, M.I., Ahmed, S., Robson, G., Javed, I., Ali, N., Atiq, N. and A. Hameed (2014). Isolation and molecular characterization of polyvinyl chloride (PVC) plastic degrading fungal isolates. *Journal of Basic Microbiology*, **54**(**1**): 18–27. https://doi.org/10.1002/jobm.201200496.

Ali, S.S., Elsamahy, T., Al-Tohamy, R., Zhu, D., Mahmoud, Y.A.G., Koutra, E., Metwally, M.A., Kornaros, M. and J. Sun (2021). Plastic wastes biodegradation: Mechanisms, challenges and future prospects. *Science of The Total Environment*, **780**: 146590. https://doi.org/10.1016/J.SCITOTENV.2021.146590.

Anwar, M.S., Kapri, A., Chaudhry, V., Mishra, A., Ansari, M.W., Souche, Y., Nautiyal, C.S., Zaidi, M.G.H. and R. Goel (2016). Response of indigenously developed bacterial consortia in progressive degradation of polyvinyl chloride. *Protoplasma*, **253**(**4**): 1023–1032. https://doi.org/10.1007/s00709-015-0855-9.

Arkatkar, A., Arutchelvi, J., Sudhakar, M., Bhaduri, S., Uppara, P.V. and M. Doble (2009). Approaches to enhance the biodegradation of polyolefins. *The Open Environmental Engineering Journal*, **2**(**1**): 68–80. https://doi.org/10.2174/1874829500902010068.

Atanasova, N., Stoitsova, S., Paunova-krasteva, T., and M. Kambourova (2021). Plastic Degradation by Extremophilic Bacteria. *International Journal of Molecular Sciences*, **22**(**11**): 5610. https://doi.org/10.3390/IJMS22115610.

Auta, H.S., Emenike, C.U., Jayanthi, B. and S.H. Fauziah (2018). Growth kinetics and biodeterioration of polypropylene microplastics by Bacillus sp. and Rhodococcus sp. isolated from mangrove sediment. *Marine Pollution Bulletin*, **127**: 15–21. https://doi.org/10.1016/j.marpolbul.2017.11.036.

Augusta, J., Müller, R. J., & Widdecke, H. (1993). A rapid evaluation plate-test for the biodegradability of plastics. *Applied Microbiology and Biotechnology* **39**(**4**): 673–678. https://doi.org/10.1007/BF00205073

Awasthi, S., Srivastava, P., Singh, P., Tiwary, D. and P.K. Mishra (2017). Biodegradation of thermally treated high-density polyethylene (HDPE) by *Klebsiella pneumoniae* CH001. *3 Biotech*, **7**(**5**): 332. https://doi.org/10.1007/s13205-017-0959-3.

Bhardwaj, H., Gupta, R., & Tiwari, A. (2013). Communities of Microbial Enzymes Associated with Biodegradation of Plastics. *Journal of Polymers and the Environment*, **21(2)**: 575–579. https://doi.org/10.1007/S10924-012-0456-Z/TABLES/1

Calabia, B.P., Tokiwa, Y., Ugwu, C.U. and S. Aiba (2010). Biodegradation. *In:* Auras, R., Lim, L.-T., Selke, S.E.M. and H. Tsuji (eds.). Poly(Lactic Acid): Synthesis, Structures, Properties, Processing, and Applications. Wiley. https://doi.org/10.1002/9780470649848.ch25.

Chamas, A., Moon, H., Zheng, J., Qiu, Y., Tabassum, T., Jang, J.H., Abu-Omar, M., Scott, S.L. and S. Suh (2020). Degradation rates of plastics in the environment. *ACS Sustainable Chemistry and Engineering*, **8(9)**: 3494–3511. https://doi.org/10.1021/ACSSUSCHEMENG.9B06635/ASSET/IMAGES/LARGE/SC9B06635_0009.JPEG.

Chen, X.Q., Guo, Z.Y., Wang, L., Yan, Z.F., Jin, C.X., Huang, Q.S., Kong, D.M., Rao, D.M. and J. Wu (2022). Directional-path modification strategy enhances PET hydrolase catalysis of plastic degradation. *Journal of Hazardous Materials*, **433**: 128816. https://doi.org/10.1016/J.JHAZMAT.2022.128816.

Cheng, H. N., Smith, P. B., & Gross, R. A. (2013). Green Polymer Chemistry: A Brief Review. In H. N. Cheng, R. A. Gross, & P. B. Smith (Eds.), ACS Symposium Series **1144**: 1–12. American Chemical Society. https://doi.org/10.1021/bk-2013-1144.ch001

Chinaglia, S., Tosin, M. and F. Degli-Innocenti (2018). Biodegradation rate of biodegradable plastics at molecular level. *Polymer Degradation and Stability*, **147**: 237–244.

Ciriminna, R. and M. Pagliaro (2020). Biodegradable and compostable plastics: A critical perspective on the dawn of their global adoption. *Chemistry Open*, **9(1)**: 8–13. https://doi.org/10.1002/OPEN.201900272.

Czajka, A., Bulski, R., Iuliano, A., Plichta, A., Mizera, K. and J. Ryszkowska (2022). Grafted lactic acid oligomers on lignocellulosic filler towards biocomposites. *Materials*, **15(1)**: 314. https://doi.org/10.3390/MA15010314.

Duru, C.E., Duru, I.A. and C.E. Enyoh (2021). In silico binding affinity analysis of microplastic compounds on PET hydrolase enzyme target of Ideonellasakaiensis. *Bulletin of the National Research Centre*, **45(1)**: 1–8. https://doi.org/10.1186/S42269-021-00563-5.

Duru, C.E., Enyoh, C.E. and I.A. Duru (2023). Juxtaposing the reactivity descriptors of plastic monomers with their binding affinity at the novel polyester hydrolase target using conceptual DFT and machine learning. *Biointerface Research in Applied Chemistry*, **13(5)**: 456. https://doi.org/10.33263/BRIAC135.456.

Duru, C., Enyoh, C., Duru, I.A. and M.C. Enedoh (2023). Degradation of PET nanoplastic oligomers at the novel PHL7 target: Insights from molecular docking and machine learning. *Journal of the Nigerian Society of Physical Sciences*, **5(1)**: 1154. https://doi.org/10.46481/jnsps.2023.1154.

Enyoh, C.E., Maduka, T.O., Duru, C.E., Osigwe, S.C., Ikpa, C.B.C. and Q. Wang (2022). In sillico binding affinity studies of microbial enzymatic degradation of plastics. *Journal of Hazardous Materials Advances*, **6**: 100076. https://doi.org/10.1016/j.hazadv.2022.100076.

Epps, T.H., Korley, L.T.J., Yan, T., Beers, K.L. and T.M. Burt (2022). Sustainability of synthetic plastics: Considerations in materials life-cycle management. *JACS Au*, **2(1)**: 3–11. https://doi.org/10.1021/JACSAU.1C00191/ASSET/IMAGES/LARGE/AU1C00191_0001.JPEG.

Faverial, J., & Sierra, J. (2014). Home composting of household biodegradable wastes under the tropical conditions of Guadeloupe (French Antilles). *Journal of Cleaner Production*, **83**: 238–244. https://doi.org/10.1016/j.jclepro.2014.07.068

Feng, S., Yue, Y., Zheng, M., Li, Y., Zhang, Q. and W. Wang (2021). IsPETase- and IsMHETase-Catalyzed Cascade Degradation Mechanism toward Polyethylene Terephthalate. *ACS Sustainable Chemistry & Engineering*, **9(29)**: 9823–9832. https://doi.org/10.1021/acssuschemeng.1c02420.

Filiciotto, L. and G. Rothenberg (2021). Biodegradable plastics: Standards, policies, and impacts. *ChemSusChem*, **14(1)**: 56–72. https://doi.org/10.1002/CSSC.202002044.

Geyer, R., Jambeck, J.R. and K.L. Law (2017). Production, use, and fate of all plastics ever made. *Science Advances*, **3(7)**: e1700782. https://doi.org/10.1126/SCIADV.1700782/SUPPL_FILE/1700782_SM.PDF.

Ghosh, S., Qureshi, A., & Purohit, H. J. (2019). Microbial degradation of plastics: Biofilms and degradation pathways. *Contaminants in Agriculture and Environment: Health Risks and Remediation*, 184–199. https://doi.org/10.26832/aesa-2019-cae-0153-014

Giacomucci, L., Raddadi, N., Soccio, M., Lotti, N. and F. Fava (2019). Polyvinyl chloride biodegradation by *Pseudomonas citronellolis* and *Bacillus flexus*. *New Biotechnology*, **52**: 35–41. https://doi.org/10.1016/J.NBT.2019.04.005.

Greene, J. (2007). Biodegradation of compostable plastics in green yard-waste compost environment. *Journal of Polymers and the Environment*, **15(4)**: 269–273. https://doi.org/10.1007/S10924-007-0068-1/TABLES/2.

Guo, B., Reddy Vanga, S., Lopez-Lorenzo, X., Saenz-Mendez, P., Rönnblad Ericsson, S., Fang, Y., Ye, X., Schriever, K., Bäckström, E., Biundo, A., Zubarev, A.R., Furó, I., Hakkarainen, M. and P.-O. Syrén (2022). Conformational selection in biocatalytic plastic degradation by PETase. *ACS Catalysis*, **12(6)**: 3397–3409. https://doi.org/10.1021/acscatal.1c05548.

Guo, J. J., Huang, X. P., Xiang, L., Wang, Y. Z., Li, Y. W., Li, H., Cai, Q. Y., Mo, C. H., & Wong, M. H. (2020). Source, migration and toxicology of microplastics in soil. *Environment International*, 137(February), 105263–105263.https://doi.org/10.1016/j.envint.2019.105263

Havstad, M.R. (2020). Biodegradable plastics. *In:* Plastic Waste and Recycling: Environmental Impact, Societal Issues, Prevention, and Solutions, pp. 97–129. https://doi.org/10.1016/B978-0-12-817880-5.00005-0.

Ho, B.T., Roberts, T.K. and S. Lucas (2018). An overview on biodegradation of polystyrene and modified polystyrene: The microbial approach. *Critical Reviews in Biotechnology*, **38(2)**: 308–320. https://doi.org/10.1080/07388551.2017.1355293.

Jambeck, J. (2015). Plastic waste inputs from land into the ocean. *Science*, **347(6223)**, 768–771. https://doi.org/10.1126/science.1260352

Jeon, H.J. and M.N. Kim (2016). Isolation of mesophilic bacterium for biodegradation of polypropylene. *International Biodeterioration & Biodegradation*, **115**: 244–249. https://doi.org/10.1016/J.IBIOD.2016.08.025.

Jerves, C., Neves, R.P.P., Ramos, M.J., da Silva, S. and A.P. Fernandes (2021). Reaction mechanism of the PET degrading enzyme PETase studied with DFT/MM molecular dynamics simulations. *ACS Catalysis*, **11(18)**: 11626–11638. https://doi.org/10.1021/acscatal.1c03700.

Jiang, R., Shang, L., Wang, R., Wang, D. and N. Wei (2023). Machine learning based prediction of enzymatic degradation of plastics using encoded protein sequence and effective feature representation. *Environmental Science & Technology Letters*, **10(7)**: 557–564. https://doi.org/10.1021/acs.estlett.3c00293.

Joo, S., Cho, I.J., Seo, H., Son, H.F., Sagong, H.-Y., Shin, T.J., Choi, S.Y., Lee, S.Y. and K.-J. Kim (2018). Structural insight into molecular mechanism of poly(ethylene terephthalate) degradation. *Nature Communications*, **9(1)**: 382. https://doi.org/10.1038/s41467-018-02881-1.

Kale, G., Anras, R., & Singh, S. P. (2006). *Comparing the degradability of commercially available biodegradable packages in composting and ambient exposure conditions.* 31–49.

Kandakatla, P., Mahto, B. and S. Goel (2013). Extent and rate of biodegradation of different organic components in municipal solid waste. *International Journal of Environment and Waste Management*, **11(4)**: 350–364. https://doi.org/10.1504/IJEWM.2013.054262.

Khan, S., Ali, S.A. and A.S. Ali (2023). Biodegradation of low density polyethylene (LDPE) by mesophilic fungus 'Penicillium citrinum' isolated from soils of plastic waste dump yard, Bhopal, India. Environmental Technology (United Kingdom), 44(15): 2300–2314. https://doi.org/10.1080/09593330.2022.2027025/SUPPL_FILE/TENT_A_2027025_SM3285.DOCX.

Kowalczyk, A., Chyc, M., Ryszka, P. and D. Latowski (2016). Achromobacterxylosoxidans as a new microorganism strain colonizing high-density polyethylene as a key step to its biodegradation. *Environmental Science and Pollution Research*, **23(11)**: 11349–11356. https://doi.org/10.1007/s11356-016-6563-y.

Kyaw, B. M., Champakalakshmi, R., Sakharkar, M. K., Lim, C. S., & Sakharkar, K. R. (2012). Biodegradation of Low Density Polythene (LDPE) by Pseudomonas Species. *Indian Journal of Microbiology, 52*(3), 411–419. https://doi.org/10.1007/s12088-012-0250-6

Leitão, A.L. and F.J. Enguita (2021). Structural insights into carboxylic polyester-degrading enzymes and their functional depolymerizing neighbors. *International Journal of Molecular Sciences*, **22(5)**: 2332. https://doi.org/10.3390/ijms22052332.

Mamin, E.A., Pantyukhov, P.V. and A.A. Olkhov (2023). Oxo-additives for polyolefin degradation: Kinetics and mechanism. *Macromol*, **3(3)**: 477–506. https://doi.org/10.3390/MACROMOL3030029.

Marin, E., Briceño, M.I. and C. Caballero-George (2013). Critical evaluation of biodegradable polymers used in nanodrugs. *International Journal of Nanomedicine*, **8**: 3071–3091. https://doi.org/10.2147/IJN.S47186.

Mihreteab, M., Stubblefield, B.A. and E.S. Gilbert (2019). Microbial bioconversion of thermally depolymerized polypropylene by *Yarrowia lipolytica* for fatty acid production. *Applied Microbiology and Biotechnology*, **103(18)**: 7729–7740. https://doi.org/10.1007/s00253-019-09999-2.

Min, K., Cuiffi, J.D. and R.T. Mathers (2020). Ranking environmental degradation trends of plastic marine debris based on physical properties and molecular structure. *Nature Communications*, **11(1)**: 727. https://doi.org/10.1038/s41467-020-14538-z.

Miyakawa, T., Mizushima, H., Ohtsuka, J., Oda, M., Kawai, F. and M. Tanokura (2015). Structural basis for the Ca^{2+}-enhanced thermostability and activity of PET-degrading cutinase-like enzyme from *Saccharomonospora viridis* AHK190. *Applied Microbiology and Biotechnology*, **99(10)**: 4297–4307. https://doi.org/10.1007/s00253-014-6272-8.

Mor, R. and A. Sivan (2008). Biofilm formation and partial biodegradation of polystyrene by the actinomycete Rhodococcusruber: Biodegradation of polystyrene. *Biodegradation*, **19(6)**: 851–858. https://doi.org/10.1007/s10532-008-9188-0.

Mohanta, T., & Goel, S. (2014). Prevalence of antibiotic-resistant bacteria in three different aquatic environments over three seasons. *Environmental Monitoring and Assessment*, **186(8)**: 5089–5100. https://doi.org/10.1007/S10661-014-3762-1/FIGURES/6

Motta, O., Proto, A., De Carlo, F., De Caro, F., Santoro, E., Brunetti, L. and M. Capunzo (2009). Utilization of chemically oxidized polystyrene as co-substrate by filamentous fungi. *International Journal of Hygiene and Environmental Health*, **212(1)**: 61–66. https://doi.org/10.1016/J.IJHEH.2007.09.014.

Muller, R (2005) Biodegradability of Polymers: Regulations and Methods for Testing: General Aspects and Special Applications, Wiley Publishers. https://doi.org/10.1002/3527600035.bpola012

Nayak, P., & Tiwari, A. (2011). Biodegradability of Polythene and Plastic By The Help of Microorganism: A Way for Brighter Future. *Journal of Environmental & Analytical Toxicology*, **1(4)**. https://doi.org/10.4172/2161-0525.1000111

Nimchua, T., Punnapayak, H. and W. Zimmermann (2007). Comparison of the hydrolysis of polyethylene terephthalate fibers by a hydrolase from *Fusarium oxysporum* LCH I and *Fusariumsolani* f. sp. pisi. *Biotechnology Journal*, **2(3)**: 361–364. https://doi.org/10.1002/biot.200600095.

Nithin, B. (2013). Biodegradation of plastic. M.Tech. thesis, IIT Kharagpur, Kharagpur, India.

Ojha, N., Pradhan, N., Singh, S., Barla, A., Shrivastava, A., Khatua, P., Rai, V. and S. Bose (2017). Evaluation of HDPE and LDPE degradation by fungus, implemented by statistical optimization. *Scientific Reports*, **7**: 39515. https://doi.org/10.1038/srep39515.

Palm, G. J., Reisky, L., Böttcher, D., Müller, H., Michels, E. A. P., Walczak, M. C., Berndt, L., Weiss, M. S., Bornscheuer, U. T., & Weber, G. (2019). Structure of the plastic-degrading Ideonella sakaiensis MHETase bound to a substrate. *Nature Communications*, **10(1)**. https://doi.org/10.1038/s41467-019-09326-3

Priya, A., Dutta, K., & Daverey, A. (2022). A comprehensive biotechnological and molecular insight into plastic degradation by microbial community. *Journal of Chemical Technology & Biotechnology*, **97(2)**: 381–390. https://doi.org/10.1002/JCTB.6675

Pischedda, A., Tosin, M. and F. Degli-Innocenti (2019). Biodegradation of plastics in soil: The effect of temperature. *Polymer Degradation and Stability*, **170**: 109017. https://doi.org/10.1016/J.POLYMDEGRADSTAB.2019.109017.

Qin, M., Chen, C., Song, B., Shen, M., Cao, W., Yang, H., Zeng, G. and J. Gong (2021). A review of biodegradable plastics to biodegradable microplastics: Another ecological threat to soil environments? *Journal of Cleaner Production*, **312**: 127816. https://doi.org/10.1016/J.JCLEPRO.2021.127816.

Ranjan, V.P. and S. Goel (2019). Degradation of low-density polyethylene film exposed to UV radiation in four environments. *Journal of Hazardous, Toxic, and Radioactive Waste*, **23**(**4**): 1–11. https://doi.org/10.1061/(ASCE)HZ.2153-5515.0000453.

Ranjan, V.P. and S. Goel (2021). Recyclability of polypropylene after exposure to four different environmental conditions. *Resources, Conservation and Recycling*, **169**: 105494. https://doi.org/10.1016/J.RESCONREC.2021.105494.

Ribitsch, D., Heumann, S., Trotscha, E., Herrero Acero, E., Greimel, K., Leber, R., Birner-Gruenberger, R., Deller, S., Eiteljoerg, I., Remler, P., Weber, T., Siegert, P., Maurer, K. H., Donelli, I., Freddi, G., Schwab, H. and G.M. Guebitz (2011). Hydrolysis of polyethyleneterephthalate by p-nitrobenzylesterase from Bacillus subtilis. *Biotechnology Progress*, **27**(**4**): 951–960. https://doi.org/10.1002/btpr.610.

Roth, C., Wei, R., Oeser, T., Then, J., Föllner, C., Zimmermann, W. and N. Sträter (2014). Structural and functional studies on a thermostable polyethylene terephthalate degrading hydrolase from *Thermobifida fusca*. *Applied Microbiology and Biotechnology*, **98**(**18**): 7815–7823. https://doi.org/10.1007/s00253-014-5672-0.

Rowe, M.D., Eyiler, E. and K.B. Walters (2016). Hydrolytic degradation of bio-based polyesters: Effect of pH and time. *Polymer Testing*, **52**: 192–199. https://doi.org/10.1016/J.POLYMERTESTING.2016.04.015.

Santacruz-Juárez, E., Buendia-Corona, R.E., Ramírez, R.E. and C. Sánchez (2021). Fungal enzymes for the degradation of polyethylene: Molecular docking simulation and biodegradation pathway proposal. *Journal of Hazardous Materials*, **411**: 125118. https://doi.org/10.1016/J.JHAZMAT.2021.125118.

Sarmah, P. and J. Rout (2019). Cyanobacterial degradation of low-density polyethylene (LDPE) by Nostoccarneum isolated from submerged polyethylene surface in domestic sewage water. *Energy, Ecology and Environment*, **4**(**5**): 240–252. https://doi.org/10.1007/s40974-019-00133-6.

Schiavo, S., Oliviero, M., Chiavarini, S. and S. Manzo (2020). Adverse effects of oxodegradable plastic leachates in freshwater environment. *Environmental Science and Pollution Research*, **27**(**8**): 8586–8595. https://doi.org/10.1007/S11356-019-07466-Z/FIGURES/4.

Shah, A.A., Hasan, F., Hameed, A. and S. Ahmed (2008). Biological degradation of plastics: A comprehensive review. *Biotechnology Advances*, **26**(**3**): 246–265. https://doi.org/10.1016/J.BIOTECHADV.2007.12.005.

Shimpi, N., Borane, M., Mishra, S. and M. Kadam (2012). Biodegradation of polystyrene (PS)-poly(lactic acid) (PLA) nanocomposites using *Pseudomonas aeruginosa*. *Macromolecular Research*, **20**(**2**): 181–187. https://doi.org/10.1007/s13233-012-0026-1.

Speight, J.G. (2019). Handbook of Petrochemical Processes. *In:* Handbook of Petrochemical Processes. CRC Press. https://doi.org/10.1201/9780429155611.

Syranidou, E., Karkanorachaki, K., Amorotti, F., Avgeropoulos, A., Kolvenbach, B., Zhou, N.Y., Fava, F., Corvini, P.F.X. and N. Kalogerakis (2019). Biodegradation of mixture of plastic films by tailored marine consortia. *Journal of Hazardous Materials*, **375**: 33–42. https://doi.org/10.1016/J.JHAZMAT.2019.04.078.

Taniguchi, I., Yoshida, S., Hiraga, K., Miyamoto, K., Kimura, Y. and K. Oda (2019). Biodegradation of PET: Current status and application aspects. *ACS Catalysis*, **9**(**5**): 4089–4105. https://doi.org/10.1021/ACSCATAL.8B05171/ASSET/IMAGES/LARGE/CS-2018-05171U_0011.JPEG.

Thomas, N.L., Clarke, J., McLauchlin, A.R. and S.G. Patrick (2015). Oxodegradable plastics: Degradation, environmental impact and recycling. *Waste and Resource Management,* **165(3)**: 133–140. https://doi.org/10.1680/WARM.11.00014.

Tokiwa, Y., Calabia, B.P., Ugwu, C.U. and S. Aiba (2009). Biodegradability of plastics. *International Journal of Molecular Sciences,* **10(9)**: 3722–3742. https://doi.org/10.3390/ijms10093722.

Tsiota, P., Karkanorachaki, K., Syranidou, E., Franchini, M. and N. Kalogerakis (2018). Proceedings of the International Conference on Microplastic Pollution in the Mediterranean Sea. pp. 181188. Available from https://doi.org/10.1007/978-3-319-71279-6.

Urbanek, A. K., Rymowicz, W., & Mirończuk, A. M. (2018). Degradation of plastics and plastic-degrading bacteria in cold marine habitats. *Applied Microbiology and Biotechnology,* **102(18)**: 7669–7678. https://doi.org/10.1007/s00253-018-9195-y

Wei, R., Oeser, T., Then, J., Kühn, N., Barth, M., Schmidt, J. and W. Zimmermann (2014). Functional characterization and structural modeling of synthetic polyester-degrading hydrolases from Thermomonosporacurvata. *AMB Express,* **4(1)**: 44. https://doi.org/10.1186/s13568-014-0044-9.

Xie, S., Sun, Z., Liu, T., Zhang, J., Li, T., Ouyang, X., Qiu, X., Luo, S., Fan, W. and H. Lin (2021). Beyond biodegradation: Chemical upcycling of poly(lactic acid) plastic waste to methyl lactate catalyzed by quaternary ammonium fluoride. *Journal of Catalysis,* **402**: 61–71. https://doi.org/10.1016/J.JCAT.2021.08.032.

Yamada-Onodera, K., Mukumoto, H., Katsuyaya, Y., Saiganji, A. and Y. Tani (2001). Degradation of polyethylene by a fungus, *Penicillium simplicissimum* YK. *Polymer Degradation and Stability,* **72(2)**: 323–327. https://doi.org/10.1016/S0141-3910(01)00027-1.

Yang, H.S., Wolcott, M.P., Kim, H.S. and H.J. Kim (2005). Thermal properties of lignocellulosic filler-thermoplastic polymer bio-composites. *Journal of Thermal Analysis and Calorimetry,* **82(1)**: 157–160. https://doi.org/10.1007/S10973-005-0857-5/METRICS.

Yang, H.S., Yoon, J.S. and M.N. Kim (2005). Dependence of biodegradability of plastics in compost on the shape of specimens. *Polymer degradation and stability,* **87(1)**: 131–135.

Yang, Y., Wang, J. and M. Xia (2020). Biodegradation and mineralization of polystyrene by plastic-eating superworms Zophobasatratus. *Science of The Total Environment,* **708**: 135233. https://doi.org/10.1016/J.SCITOTENV.2019.135233.

Yoshida, S., Hiraga, K., Taniguchi, I. and K. Oda (2021). Ideonellasakaiensis, PETase, and MHETase: From identification of microbial PET degradation to enzyme characterization. *In:* Gert Weber, G., Bornscheuer, U.T. and R. Wei (eds.). Methods in Enzymology. Academic Press, **648**: 187–205. https://doi.org/10.1016/bs.mie.2020.12.007.

Yue, W., Yin, C.F., Sun, L., Zhang, J., Xu, Y. and N.Y. Zhou (2021). Biodegradation of bisphenol-A polycarbonate plastic by *Pseudoxanthomonas* sp. strain NyZ600. *Journal of Hazardous Materials,* **416**: 125775. https://doi.org/10.1016/J.JHAZMAT.2021.125775.

Zhang, H., Hou, Y., Zhao, W. and H. Na (2022). Control strategies of plastic biodegradation through adjusting additives ratios using in silico approaches associated with proportional factorial experimental design. *International Journal of Environmental Research and Public Health,* **19(9)**: 5670. https://doi.org/10.3390/ijerph19095670.

Zhang, J., Gao, D., Li, Q., Zhao, Y., Li, L., Lin, H., Bi, Q. and Y. Zhao (2020). Biodegradation of polyethylene microplastic particles by the fungus Aspergillus flavus from the guts of wax moth, *Galleria mellonella. Science of The Total Environment,* **704**: 135931. https://doi.org/10.1016/J.SCITOTENV.2019.135931.

Zhang, K., Hamidian, H., Tubi, A., Zhang, Y., Fang, J. K. H., Wu, C. and P.K.S. Lam (2021). Understanding plastic degradation and microplastic formation in the environment: A review. *Environmental Pollution,* **274**: 116554. https://doi.org/10.1016/j.envpol.2021.116554.

Zhou, Y., Kumar, M., Sarsaiya, S., Sirohi, R., Awasthi, S. K., Sindhu, R., Binod, P., Pandey, A., Bolan, N. S., Zhang, Z., Kumar, S., & Awasthi, M. K. (2022). Challenges and opportunities in bioremediation of micro-nano plastics: A review. *Science of the Total Environment,* **802**. https://doi.org/10.1016/j.scitotenv.2021.149823

Zhu, B., Wang, D. and N. Wei (2022). Enzyme discovery and engineering for sustainable plastic recycling. *Trends in Biotechnology,* **40(1)**: 22–37. https://doi.org/10.1016/j.tibtech.2021.02.008.

Index

S. Goel (ed.), *Advances in Solid and Hazardous Waste Management*,
https://doi.org/10.1007/978-3-031-49144-3

Q

R

SPRINGER NATURE

GPSR Compliance

The European Union's (EU) General Product Safety Regulation (GPSR) is a set of rules that requires consumer products to be safe and our obligations to ensure this.

If you have any concerns about our products, you can contact us on ProductSafety@springernature.com

In case Publisher is established outside the EU, the EU authorized representative is:

Springer Nature Customer Service Center GmbH
Europaplatz 3
69115 Heidelberg, Germany

The manufacturer's authorised representative in the EU is Springer Nature Customer Service Centre GmbH, Europaplatz 3, 69115 Heidelberg, Germany. If you have any concerns regarding our products, please contact ProductSafety@springernature.com

Printed and bound by CPI Group (UK) Ltd, Croydon, CR0 4YY
01/10/2026
02225935-0003